WATER DEVELOPMENT, SUPPLY AND MANAGEMENT

Series Editor: ASIT K. BISWAS

Volume 6

INTERREGIONAL WATER TRANSFERS

WATER DEVELOPMENT, SUPPLY AND MANAGEMENT

Series Editor: ASIT K. BISWAS

Other titles in the series:

Volume 1	UNITED NATIONS	Water Development and Management: Proceedings of the United Nations Water Conference, in 4 volumes
Volume 2	BISWAS, ASIT K.	United Nations Water Conference: Summary and Main Documents
Volume 3	WORTHINGTON, E.B.	Arid Land Irrigation in Developing Countries: Environmental Problems and Effects
Volume 4	PACEY, A.	Water for the Thousand Millions
Volume 5	UNITED NATIONS	Register of International Rivers
Volume 7	WIDSTRAND, C.	Water and Society: Conflicts in Development Part 1: The Social and Ecological Effects of Water Development in Developing Countries
Volume 8	WIDSTRAND, C.	Water and Society: Conflicts in Development Part 2 - Water Resources, Needs and Conflicts
Volume 9	OUANO, E.A.R.	Water Pollution Control in Developing Countries
Volume 10	HUSSEINY, A.A.	Iceberg Utilization
Volume 11	INTERNATIONAL COMMISSION ON IRRIGATION AND DRAINAGE	The Application of Systems Analysis to Problems of Irrigation, Drainage and Flood Control

INTERREGIONAL WATER TRANSFERS

Problems and Prospects

Editors

GENADY N. GOLUBEV
International Institute for Applied Systems Analysis, Laxenburg, Austria

ASIT K. BISWAS
Director, Biswas & Associates, 3 Valley View Road, Ottawa, Canada, and
International Institute for Applied Systems Analysis, Laxenburg, Austria

PERGAMON PRESS

OXFORD · NEW YORK · TORONTO · SYDNEY · PARIS · FRANKFURT

U.K.	**Pergamon Press Ltd., Headington Hill Hall, Oxford OX3 0BW, England**
U.S.A.	**Pergamon Press Inc., Maxwell House, Fairview Park, Elmsford, New York 10523, U.S.A.**
CANADA	Pergamon of Canada, Suite 104, 150 Consumers Road, Willowdale, Ontario M2J1P9, Canada
AUSTRALIA	Pergamon Press (Aust.) Pty. Ltd., P.O. Box 544, Potts Point, N.S.W. 2011, Australia
FRANCE	Pergamon Press SARL, 24 rue des Ecoles, 75240 Paris, Cedex 05, France
FEDERAL REPUBLIC OF GERMANY	Pergamon Press GmbH, 6242 Kronberg-Taunus, Pferdstrasse 1, Federal Republic of Germany

First edition 1979

British Library Cataloguing in Publication Data

International Institute for Applied Systems
Analysis. Task Force Meeting, *Laxenburg, 1977*
Interregional water transfers. - (Water
development, supply and management; Vol. 6).
1. Water transfer - Congresses
I. Title II. Golubev, Genady N
III. Biswas, Asit K IV. Series
331.9'1 HD1691 78-40769

ISBN 0-08-022430-X

Also published as volume 2 number 2 of journal *Water Supply and Management* and supplied to subscribers as part of their 1978 subscription.

Views expressed herein are those of the contributors and not necessarily those of the International Institute for Applied Systems Analysis.

Printed in Great Britain by A. Wheaton & Co. Ltd., Exeter

Contents

Preface

The total amount of water available globally, if used more efficiently, can meet vastly higher human needs. Current estimates indicate that the total volume of water on earth is 1.4×10^9 km^3, 97.3 percent of which is ocean water, and, therefore cannot be used by man except for fisheries and navigation. Only 2.7 percent is fresh water, 77.2 percent of which is stored in polar ice-caps and glaciers, 22.4 percent as ground water and soil moisture (about two-third lies deeper than 750 metres below the surface), 0.35 percent in lakes and swamps, 0.04 percent in the atmosphere and less than 0.01 percent is in streams. In other words, nearly 90 percent of fresh water is stored in ice-caps, glaciers and as deep ground water, and as such is not easily accessible. For all practical purposes, it is surface water in rivers, streams and lakes, amounting to less than half of 1 percent of available freshwater, that constitutes the basic available supply for man, even though ground water has been heavily developed in certain parts of the world.

With the continual increase in the world population, more and more water is becoming necessary for domestic, industrial, agricultural and other purposes. Furthermore, it has been observed that as the standard of living increases, water use per capita goes up as well. Thus, mankind is going to need more and more water in the future, both due to increases in population and standard of living.

The major user of water, however, is agriculture. It has been estimated that agriculture accounts for 80 percent of all water consumption in the world: the corresponding figure for the United States is slightly above 40 percent. It takes approximately 1,000 tons of water to grow one ton of grain and 2,000 tons to grow one ton of rice. In addition, animal husbandry and fisheries require abundant water.

In 1975, according to the Food and Agricultural Organization of the United Nations, the total area irrigated in the world amounted to 223 million hectares, of which 93 million hectares was in developing countries. Some 15 per cent of the world's cropland is irrigated, but it yields from 30 to 40 per cent of all agricultural production. The amount of water used by irrigated crops is nearly 1,300,000 million cubic metres, but because of losses in storage, conveyance and use, the total amount used increases to almost 3,000,000 million cubic metres.

It is estimated that by 1990 the total area irrigated in the world will increase to 223 million hectares, of which 119 million hectares will be in developing countries. Expanding and maintaining irrigated areas to 1990 is going to be a challenging task, and its magnitude can be judged by the following requirements for the developing market economy countries only:

22.5 million hectares of new irrigation;

45 million hectares of irrigation improvement;

78.2 million hectares of drainage improvement, including 52.4 million hectares on irrigated land;

438,000 million m^3 of additional water; and $97,800 million of investment at 1975 prices.

Increased agricultural activities in marginal areas have often overexploited water availability. In many areas, more groundwater is being withdrawn than can be replenished naturally, thus contributing to major management and environmental problems.

As more and more water becomes necessary for agricultural, industrial and other purposes, certain regions of the world have already started to face shortages, so much so that it is becoming one of the major constraints for further socioeconomic development. Accordingly, it is necessary to explore alternatives to alleviate this rapidly developing critical situation. One possibility is to transfer water over long distances from surplus to deficit areas. Already many projects exist which divert water from one region to another.

In order to study the different aspects of interregional water transfer, Resources and Environment Area of the International Institute for Applied Systems Analysis (IIASA) Laxenburg, Austria, convened a Task Force meeting of international experts from Canada, India, Mexico, Soviet Union and the United States. This meeting was held at IIASA during October 11-14, 1977. It discussed not only the experiences of different countries in interregional water transfers, but also their geophysical, environmental and economic implications. The papers included in this volume are the results of that meeting. The only new addition has been a comprehensive bibliography on the subject.

Genady Golubev
International Institute
for Applied Systems Analysis,
Laxenburg,
Austria.

Asit K. Biswas, Director
Biswas & Associates
3 Valley Road, Ottawa, Canada,
and
International Institute for Applied
Systems Analysis,
Laxenburg,
Austria.

Water Supply & Management, Vol. 2, pp.59-65. 0364-7714/78/0601-0059$02.00/0

Interregional Water Transfers

GENADY GOLUBEV* and ASIT K. BISWAS**

A Task Force, comprising well-known international experts, met at the International Institute of Applied Systems Analysis (IIASA), Laxenburg, Austria, during 11–14 October, 1977, to discuss and review current status of interregional water transfers (IWT) in the world, and to make some recommendations about possible future directions of work. Specialists on IWT from Canada, India, Mexico, Soviet Union and the United States attended the meeting under the chairmanship of Professor Genady Golubev of IIASA; Dr. Asit K. Biswas of Canada was the General Rapporteur.

In his opening address, Dr. Roger Levien, Director of IIASA, briefly described the current research of the Institute, and stressed the importance of IWT within the framework of the existing research activities in the area of water.

The Chairman of the meeting, Professor Golubev, then set the scene for the 4-day meeting by raising some principal questions with regard to IWT projects. He pointed out five major considerations. These were:

(1) The size of IWT projects has been growing exponentially with respect to time. Now the largest ones can transfer up to 10 km^3/yr. Projects for the next 20–30 years are of the next order of magnitude.

(2) Some groups of problems arise because of the growing size of IWT projects: (a) water demand/supply relationships as a starting point for IWT; (b) uncertainty; (c) efficiency; (d) links with other major problems (energy, resources, capital investment, food, etc.); (e) impacts; and (f) other, non-conventional ways of water supply.

(3) In the USSR and the USA, IWT projects have stemmed from: (a) serious demand/supply situations in southern USSR and south-western USA; (b) decrease of river run-off due to human activity; and (c) deterioration of hydrologic regimes of lakes and seas.

(4) The IWT problem consists of three main blocks: technology, socio-economic, and environment. They are subdivided into sub-blocks of a lower level. There is a strong interrelation not only *within* the main blocks but also *between* them.

(5) As a general rule, as the size of IWT projects increases, the complexity increases as well. Uncertainty is in turn connected with complexity and is also growing. Comparison of the curve of uncertainty and the curve of efficiency as functions of the projects' size has been demonstrated. With these considerations in mind, it can be concluded that there is a certain size limitation above which uncertainty is greater than efficiency, and very big IWT projects are not appropriate for the time-being. With the progress of science the critical size of projects will increase.

Dr. Asit K. Biswas, Director of Biswas & Associates, Ottawa, Canada, provided an overview of the interregional water transfer projects in North America. He pointed out that three

* International Institute for Applied Systems Analysis, Laxenburg, Austria.

** Director, Biswas & Associates, 3 Valley View Road, Ottawa, Canada, and International Institute for Applied Systems Analysis, Laxenburg, Austria.

factors must be analyzed before IWT could be considered. These are availability of water, both in terms of space and time, nature of demand functions and current efficiency of water use. In many parts of North America, especially in northern Canada and parts of Mexico, adequate data on surface and groundwater supplies do not exist. In many other parts of Canada, the United States and Mexico data are available only for a short period of time and hence reliable forecasts of water availability on a probabilistic basis are difficult to make. The situation is much worse when water demands are considered. Demand functions are difficult to construct, and in the context of water planning demands are often synonymous with requirements. Finally, efficiency of water use is very low in certain sectors, especially in agriculture. On a global basis, 80% of total water used is for agriculture: the corresponding figure for the United States is about 40%. Currently, 223 million ha of land are irrigated in the world, 93 million ha of which are in developing countries. Irrigated crops currently require 1.3 million million m^3 of water, but because of losses in distribution systems, 3 million million m^3 of water have to be withdrawn. The efficiency of global irrigation is even much less since there is a universal tendency to over-irrigate. Thus, in most cases, before major IWT schemes can be considered, it would make better sense to improve the water use efficiency of present systems.

The most ambitious IWT plan in North America was the North American Water and Power Alliance (NAWAPA), first proposed in 1964. The general approach of this scheme was to distribute the surplus water of the high precipitation areas of the north-western part of the North America to water scarce areas of Canada, the United States and Mexico. The immensity of the plan stirred the imagination of many engineers and economists, and within the 5 years of NAWAPA being proposed, a whole series of IWT schemes was put forward to redistribute the waters of North America.

However, as these new massive diversion schemes were being proposed, a new era dawned in North America. Toward the end of the 1960s, environmental considerations became increasingly more important, and this culminated in the development of a completely new process – that of environmental impact assessment – within the overall planning framework. Politically, environmentalists became a major force, and they opposed construction of massive water development projects on environmental and ecological grounds. The growth of environmental awareness, to a large extent, contributed to the decline of interest in IWT in Canada and the United States. At the present time, it is hard to foresee the construction of any new major IWT in Canada and the United States before the end of this century.

Prof. G. Voropaev, Director of the Water Problems Institute, Moscow, reviewed the Soviet experiences in IWT. The long-term economic planning in the USSR foresees considerable growth of water demands. By the end of the present century, water demands will exceed the present level by two to three times. The existing resources will not be enough to meet the growing water demands in the southern parts of the USSR. To meet these demands, it is necessary to undertake complex measures that will include the following:

(1) the improvement of the technology of the water use and the substitution of water-consuming industries by less consuming ones;

(2) fuller use and the increase of water supply from local water resources by run-off regulation;

(3) the territorial redistribution of water resources by redirecting run-off of the northern rivers to the southern side.

The most important matter in solving the problem of water needs of the national economy will be the territorial redistribution of water resources. The choice of the alternatives and the sequence in taking measures on the territorial redistribution are possible only by indepth study of the problem. Such a study will provide predictions of the long-term impacts on ecological, physio-geographical, and socio-economical processes by water redistribution measures. It is critical to realize the interrelations between these processes in order to understand their regional estimation, to study the dynamics of their development, and to see the global aspects of the problem. Studies of this kind have already been initiated in the USSR. Their methodological foundation is based on the systems approach to the problem.

The complexity of this problem solution is also conditioned by a number of specific factors such as a wide range of climatic changes over the vast territory of the USSR, extremely uneven distribution of surface and groundwater resources, the existence of large water bodies (seas) in the south, synchronous or asynchronous river run-off oscillations over big territories of the country, water demands in various regions, etc.

At the same time in the USSR there are a number of objective prerequisites for successful solution of this problem, two of which are: general state planning of the whole economic and social life of the society and people's property of land and water resources; and the high economic power of the country and large experience in conducting large-scale water projects on irrigation, hydroenergetics and water transport.

H. Garduño of the Comisión del Plan Nacional Hidráulico, Mexico City, described the current plans for large-scale transfers within the master water plan for Mexico. Water resources planning in Mexico is carried on by the National Water Plan Commission (NWPC), from the Agriculture and Water Resources Ministry. In a 5-year period, a special planning process was designed and the National Water Plan (NWP) 1975 was completed.

The methodology consists of an iterative process with both national and regional approaches. Each iteration starts with alternative socio-economic scenarios and its main results are national and regional objectives, goals, policies and programs for each basic (e.g. irrigation, flood control, etc.) and supportive activity (e.g. research, water inventories, etc.).

It is within this context that the need for water transfers appears. In Mexico, a country of 200 million ha, with a population of 60 million in 1975 and a mean annual run-off of 410 km^3, agricultural soil, water and population are unevenly distributed and they do not coincide geographically. The total irrigated area is 5 million ha, 900,000 of which lie in the northwestern regions, where there still is a surface of 1.5 million ha of good idle land. Eighty per cent of this surface is located in the northern part, while the rivers which are still uncontrolled lie in the south. To irrigate about 900,000 new ha by the turn of the century, a combination of aquifer mining during 10 years and water transfers to irrigate new lands and to rescue the lands irrigated with mined groundwater will be developed. The system will include the construction of eight dams, conduits of 1500 km and some 600 Wh per year of energy to raise the water 500 m.

The NWPC is presently working on linking some models it has developed during the last 3 years so that, once a national goal is set up by the government (e.g. food self-sufficiency) the evaluation procedures help to decide which projects are better to achieve that goal.

The other water transfer project for the near future is needed to supply water for Mexico with a present population of more than 10 million. The supply will have to increase from 42 m^3/sec to 110 m^3/sec in the year 2000, with pumping heads of more than 1000 m to reach the Mexican capital at 2240 m above sea level. The huge investments needed, and the population estimates (by the year 2000, 30 million) make it clear that higher efficiency is needed in water use and that effective decentralization measures should be taken to reach a more balanced regional development in the country.

Robin R. Reynolds, Deputy Director of the California Department of Water Resources, Sacramento, reviewed the Californian experience in the operation of IWT projects.

Using California and California's State Water Project as an example, the phases in the history of the development and use of water resources were reviewed. Several of the examples of systems analysis used in the planning and operation of the Project were reviewed, especially those relating to operations in the Sacramento–San Joaquin Delta where there is intense technical and political controversy. The operation scheme using off-peak power also was described. In addition, a possible ultimate pattern of water development for a nation or a large international region was discussed. It was suggested that some insight into the characteristics of one such possibility of an ultimate phase can be gained by considering the characteristics of a large power grid system. On this basis, the characteristics of a water grid were described. The most significant characteristics are large interbasin and interregional aqueducts and a central coordination and management.

Mr. K.S. Murthy of the Central Water Commission of the Government of India, New Delhi, India, reviewed the current status and future plans of IWT in India, a country that has a geographical area of over 800 million acres. The cropped area is about 400 million acres. Current irrigation covers nearly 100 million acres. The ultimate irrigation potential is estimated at over 200 million acres.

India lies in the tropical and sub-tropical region. Rainfall is confined to the monsoon months of June–September (nearly 90%). It is erratic – not dependable in most parts of the country. Agriculture is the main occupation of the people – over 70% are engaged in it. It contributes over 50% of the GNP but successful agriculture is not possible in most areas of the country without irrigation.

Since Independence (1947) the country has embarked on a massive irrigation development program. Over 70 million rupees have been spent so far. Current annual investment is over 10 billion rupees. One third of the country suffers from drought. Large sums of money are spent in relief works. To provide permanent relief, studies and investigations are now in progress for big irrigation projects for these areas. These involve interregional interbasin transfer of water. Obviously, drought areas have no waters of their own.

Under the Constitution of India, "water" is under state jurisdiction. The Central Government acts as a coordinator and provides technical and financial assistance, and in certain cases helps in construction as well. Inter-State agreements are necessary for interbasin and interregional water transfers, which take time. Proposals are under consideration to give the Centre greater authority in this matter.

Inter-linking of rivers has been under consideration for quite a few years – north to south, east to west, etc. This has generated a lot of passion and arguments. Current studies envisage interregional water transfer taking note of local needs and sentiments. But at the same time these are being so designed as to fit into an over-all national water grid at a future date. The main elements of such transfers are high-lift storage and long-distance movement.

Agronomic and economic aspects are equally important, especially for high-lift water uses. Political considerations should be given due consideration and public opinion is important for construction of these schemes. Once public opinion develops to support the schemes, the task becomes easier to accomplish.

Interregional water transfers are going to be crucial in the coming years. They are the answer to the "Two Faces of Water – Floods and Droughts". International cooperation, especially in the fields of shared knowledge and experience, can play a vital role in this field.

Professor Charles W. Howe of the Department of Economics, University of Colorado, Boulder, Colorado, reviewed the history of IWT in the USA, which is similar to other countries in its progression from smaller to larger projects. The high costs, uncertain environmental effects, and opposition from areas of origin led to reduced interest in IWT until the energy crisis of 1973. Some interest has arisen from potential energy industry demands such as shale oil production and coal gasification. Other current interest arises from "rescue operations" for regions dependent on fossil groundwater.

Six issues warrant discussion. (1) Agriculture is generally the largest user of proposed transfers but represents the lowest valued uses, reducing benefit–cost ratios. At the same time, low agricultural values imply the possibility of satisfying new demands with present sources of agriculture water rather than from IWT. (2) Limited world markets study the price effects of large expansions of irrigated production. (3) Possibilities of increasing use efficiency of existing supplies as a substitute for IWT or to defer the need for IWT construction must be studied. (4) The extent of energy recovery in IWTs which require pumping is important to economic feasibility. (5) Externalities in the exporting region such as foregone uses and increased salinity concentrations must be taken into account. (6) There is a tendency toward premature construction of IWTs. Several case studies (Arizona and Mexico) have shown that deferring IWT projects would not be costly to the importing region and would greatly reduce the present value of construction costs. These last two points relate closely to the major points of Professor Fisher.

Professor Anthony Fisher of the University of California, Berkeley, reviewed some theoretical and measurement problems in economic assessment of IWT projects. With recent increases in the size of proposed IWTs, careful consideration of their economics becomes particularly important. A number of proposals and propositions concerning the theory and measurement of the costs and benefits of an IWT were suggested. They can be stated briefly as follows:

(1) Commonly used methods of measuring the conventional economic impact, including input–output analysis, are not entirely adequate, in that they do not allow sufficiently for induced changes in the structure of the economies of the impacted regions, do not trace these changes through time, and do not relate them to maximizing behavior by economic agents. An econometric modeling approach may be used to accomplish these objectives.

(2) Calculations of the benefits and costs of an IWT ordinarily ignore its effects on the environment, yet these are likely to be substantial. The standard decision criterion can be modified to include the costs of environmental effects. Where the costs cannot be estimated, a technique was suggested for comparing an IWT to an alternative means of producing water, that still accounts for both conventional economic and environmental effects of each. Briefly, if both the economic and the environmental costs (even where these cannot be measured in money terms) are lower for one of the alternatives, it is said to *dominate*.

(3) It is possible that the environmental effects of an IWT may be both irreversible and uncertain. Where, however, the uncertainty diminishes over time, as better information about the effects and their costs becomes available, there is a kind of additional cost to proceeding "too soon" with the project. This represents a further modification of the standard benefit–cost criterion.

Discussion of the paper was brief, limited mainly to technical questions about part (3). The one important point of substance had to do with the practicality of the finding in part (3). That is, would it be practical to estimate the additional cost to proceeding "too soon"? Fisher's answer was, probably not; but the finding is still relevant, since it puts the burden of proof on marginal projects. A project that exhibits the characteristics of part (3) should not be undertaken (on economic efficiency grounds) if its expected present value is just barely positive.

Professor Leonard Ortolano of Stanford University, USA, discussed environmental assessments in water resources planning, with special reference to the United States. As a result of laws and regulations promulgated early in the 1970s, environmental assessments are required for studies carried out by the federal agencies responsible to water resources planning in the United States; a wide variety of impacts on both the natural and "social" environment are included in these assessments. The paper summarizes the nature of these impacts as well as the various methods being used for impact identification, prediction and evaluation. Documentation from case studies and mail questionnaire surveys is presented to support the notion that issues related to assessment methods *per se* are not the critical ones in ensuring that environmental factors receive adequate consideration in planning. Rather, the key issues relate to the ways in which the results from applying environmental assessment methods are used in water resources planning and decision making. An "iterative, open-planning process" is presented as providing a mechanism for assuring that environmental assessments are used in the formulation and ranking of alternative actions; such a process is now being used by the US Army Corps of Engineers, one of the principal water resources agencies in the United States. The discussion of the paper indicated that the use of an iterative, open-planning process in no way detracts from the efforts of economists to evaluate environmental effects in monetary terms and thereby increase the extent to which alternatives are evaluated on a rigorous, systematic basis.

Professor G. Golubev reviewed environmental issues of big IWT projects. With the increase of the size of IWT projects the complexity of environmental assessment is growing as well. It stems from the fact that the number of components and links in a geoecosystem are increasingly nonlinearly as a function of projects' size. Correspondingly, the greater the size, the greater the uncertainty in evaluation of environmental impacts. Above a certain size, uncertainty would be so high that it would not be feasible to carry out a project. Possibly it is true of very big IWT projects regarding present knowledge of environmental assessment. One of the approaches to decrease uncertainty may be the dividing of a complex project into parts.

A case study of environmental issues of big IWT projects has been done by the author using one of the proposals for reallocation of water in the USSR. (The mouth of the Ob River, Ural Mountains, Pechora River, Severnaya Dvina River, Volga River, Central Asia and Don River.) All this long way was divided by reaches. Environmental problems concerning IWT were expressed in the form of trees of consequences, or networks. The main problems have been discussed for each reach (Arctic Sea problem, change of regime in rivers, change of adjacent territories, impoundments and related effects, improvement of hydrologic regime of the Azov Sea, development of irrigation and related problems, etc.). The approach is regarded as being useful both for the first steps in the project's assessment and for better management of scientific studies for environmental forecast of IWT projects.

CONCLUSIONS AND RECOMMENDATIONS

The Task Force agreed on several conclusions. These are the following:

(1) Interregional water transfers are and will be one of the ways to increase water supply. However, they should not be regarded as the only means available to an end: rather it should be considered as one of several alternatives available to optimize water use. If, having analyzed

all these alternatives, IWT appears as a promising solution, it should be considered within the planning framework. The timing of an IWT project should be elaborated having in mind other alternatives of water supply.

(2) To assess a large IWT project it is not sufficient to apply conventional methods of an economic evaluation and much more broad, complex approach is required.

(3) The environmental and ecological costs of IWT could be many, and these should be carefully analyzed and evaluated. Adequate counter-measures must be taken to reduce such costs to a minimum. The least solved question within the IWT problem is a methodology to assess environmental costs of IWT's and to forecast their impacts on nature.

(4) Similarly, social costs of such schemes should also be evaluated.

(5) The feasibility of IWT varies from country to country and region to region. In other words, it is site-specific. Whereas it is unlikely that new IWT schemes could be developed in the United States and Canada because of economic, environmental and political reasons, at least within the next two decades, it seems that they are viable under certain situations in countries like the Soviet Union, India or Mexico.

The Task Force made the following recommendations:

(1) The papers prepared for the Task Force meeting should be published as soon as possible, since these provide an authoritative account of the present status of IWT in the world on an interdisciplinary basis. No such comparable work is currently available. It was agreed that Professor Genady Golubev and Dr. Asit K. Biswas will edit the proceedings, which would be published in both the English and Russian languages.

(2) IWT should be considered within the over-all context of other non-conventional means of water development like weather modification, iceberg towing, desalination, use of very large crude carriers (VLCC) to transport fresh waters, etc. An over-all state-of-the-art report, critically reviewing the present developments in such areas, is both necessary and desirable.

(3) Within the context of IIASA's present program on water demand modeling, IWT should be considered as one of several other alternatives, wherever desirable and feasible. Attempts should be made to develop general guide-lines for IWT and these then should be incorporated within a methodological context.

(4) IIASA can play a catalytic role in IWT, and the Institute can play a major role in terms of information exchange, on a global basis, in this field.

(5) It was suggested that another meeting be held in about 2 years' time. As far as there is no developed methodology to assess and forecast environmental implications of current and proposed IWT schemes, future meeting should consider this issue. Critical state-of-the-art reviews on the subject would be desirable.

If necessary, small *ad hoc* teams should be set up – comprising IIASA and external specialists – on specific sub-topics.

Water Supply & Management, Vol. 2, pp.67-77. 0364-7714/78/0601-0067$02.00/0

Interregional Water Transfers as an Interdisciplinary Problem

GENADY GOLUBEV* and OLEG VASILIEV*

Interregional water transfer is a fairly popular and important topic. For example in the Soviet Union, about 100 institutes are working on the problem of reallocating water resources of the country. Recently, the same issue was also discussed in the United States. There are large water transfers projects in India and Mexico, and plans, projects and even some construction are already underway in Australia, Pakistan, Hungary, Sudan, Egypt, Canada and other countries.

The topic of water transfers is not a new one. Water diversions are known from former times in the States of Ur, Rome and by the Incas. In the last century, especially in the last few decades, many water transfers have been carried out. In Tables 1 and 2, some features of the big water transfer schemes in the US and USSR are given. Some of the values are perhaps out-of-date, as some projects continue to develop (i.e. Karacum Canal). The total amount of reallocated water in the USSR, is now about 50 km^3/yr. The same order of magnitude is characteristic of the US.

Throughout the history of water transfers, their sizes (such as water discharge, length of canal, etc.) have grown exponentially. At the beginning of this century, water discharges from the largest transfer schemes were about 0.5 or 1 km^3/yr. They are currently of the next order of magnitude (see Tables 1 and 2, compiled by the authors) and new projects of another order of magnitude are proposed. The situation is represented graphically in Fig. 1.

Some groups of problems arise because of the growing size of water transfer projects, for example:

(1) water demand/supply relations as a starting point for decision making concerning water transfers;

(2) uncertainty of large interregional water transfer (IWT) projects and control of them;

(3) efficiency (not only expressed in terms of money);

(4) links with other major problems of global or universal interest (energy, resources, food, capital investments, etc.);

(5) consequences and impacts (both short-term and long-term ones); and

(6) other water supply alternatives to IWT.

Let us look into some of these questions.

Water demand/supply relations in a number of regions or countries are rather unfavorable. Table 3 gives an idea of water resources for a year of low flow (75% probability) in comparison with water demands for eight major rivers of the southern European part of the USSR.[1] It is obvious that every fourth year the situation is rather strained and if we take into account constant growth of water demand, especially the projected considerable increase of irrigated lands, the supply/demand difference would be more unfavorable.

---

* International Institute for Applied Systems Analysis, Laxenburg, Austria.

Table 1. Some water transfers in the USA

Name of the project	State	Direction of the transfer	Amount of transferred water in km^3/yr	Total length in km		Water reservoirs		Total capacity of hydropower stations in 10^3kWt	Additional data
				Canals	Tunnels	Number	Total volume in km^3		
Colorado Big Thompson (1938 - 1959)	Colorado	Colorado River, Rocky Mts, South Platte River basin	0.4	154	55	10	1.2	180	4 pumping stations with total capacity of 30×10^3kWt
Central Arizona (1968 - 1985)	Arizona	Colorado River, Central Arizona	2.7	600	10	4	?		9 pumping stations with total capacity of 547×10^3kWt
Fryingpan - Arkansas (1962 -)	Colorado	Colorado River basin Arkansas River basin	0.1	?	42	4	0.9	200	
Central Utah (1964 -)	Utah	Between river basins in the state of Utah	0.2	190	10	6	1.3	133	
California State water project	California	Reallocation of water in state (mainly from North to South)	4.2 (1972) 5.5 (plan)	1100	?	23	8.2	630	18 pumping stations
Central Valley Project	California	Reallocation of water in the state (mainly from North to South)	9.0	?	-	19	13.8 (1972) 19.6 (plan)	1250 (1972) 1820 (plan)	
All systems of California	California	Reallocation of water in the state (mainly from North to South)	31.1	?	?	68	about 30	?	

Table 2. Principal water transfers in the USSR

Donor river	Name of canal	Water discharge km³/yr	Length km	Principal use of water
		Interbasin transfers		
Volga	Volga - Moscow	2.3	100	Municipal and industrial
Amudaria	Karacum	7.8	760	Agricultural
Dnieper	North Crimea	8.2	400	Agricultural
Irtysh	Irtysh - Karaganda	2.2	460	Industrial
Samur	Samur - Apsheron	1.7		Agricultural and industrial
Dnieper	Dnieper - Donbass	1.2		Industrial
Volga	Volga - Ural	3.1	400	Agricultural
		Intrabasin transfers		
Naryn	Great Fergana	6.0	350	Agricultural
Syrdaria	Golodnaya Step	4.4		Agricultural
Kura	Verhnii Karabakh	3.6	170	Agricultural
Kuban	Nevinnomysskii	1.9	50	Agricultural
Don	Don Magistralnii	1.0	110	Agricultural
Terek	Tersko - Kumskii	2.7	150	Agricultural
Kura	Verhnii Shirvanskii	2.4	120	Agricultural

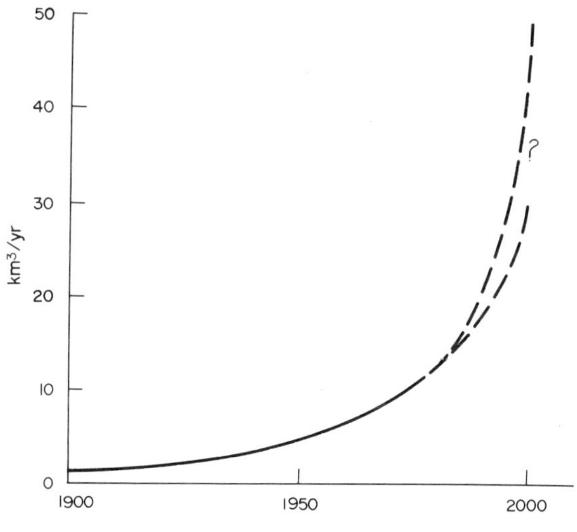

Fig. 1. Size of water transfer projects as a function of time.

In the US, the water resources of the western and southwestern areas are considerably below those of the rest of the country. Table 4 gives an idea of resources and demand differences for this part of the US (data taken from Howe and Easter[2]). Tables 3 and 4 can not be compared directly; to the values in Table 4 on-stream use and sanitary run-off should be added, thus increasing the water demand 30 or 40% (these values have been obtained from Soviet data). According to Table 4, with corrections, the situation is quite strained. Note that water surplus in the Pacific Northwest is fictitious, as hydropower stations there require a greater proportion of on-stream use of water.

Table 3. Comparison of water supply and water demand for major rivers of the southern European part of the USSR for a low flow year (75 % level of probability), km³/yr (*after*

River Basin	Supply (river run-off plus underground water)	Demand							Surplus (+) or Deficit (-)
		Consumptive use					Instream uses (sanitation, navigation, electrical power generation, etc.)	Total water demand	
		Irrigation and other agricultural needs	Municipal and industrial	Fish-breeding	Evaporation from water reservoirs	Total consumptive use			
Volga	221.5	15.6	4.4	2.9	19.2	42.1	168.0	210.1	+11.4
Dnieper	45.4	14.7	6.7	0.6	3.2	25.2	16.0	41.2	+4.2
Kura	25.5	13.9	0.8	0.7	2.5	17.9	5.5	23.4	+2.1
Don	21.7	6.0	2.0	0.7	2.0	10.7	21.0	31.7	-10.0
Kuban	12.0	7.2	0.4	2.3	0.2	10.1	2.0	12.1	-0.1
Terek and Sulak	15.4	6.4	0.7	1.9	0.0	9.0	4.5	13.5	+1.9
Dniester	8.3	2.6	0.8	0.4	0.2	4.0	2.5	6.5	+1.8
Ural	5.3	1.1	1.1	0.8	0.9	3.9	5.5	9.4	-4.1

Table 4. Comparison of water supply and water demand for the western part of the USA, 1965 (Howe and Easter[2])

Water use region	run-off		Consumptive use				Surplus (+) or deficit (-) as to:	
	Average	90% of probability	Irrigation and other agricultural needs	Municipal and industrial	Conveyance losses	Total	Average Run-off	Low run-off ($P = 90\%$)
Western Gulf	60	23.4	18.7	1.7	4.6	25.0	+35	-1.6
Colorado	23	12.4	11.1	0.4	4.2	15.7	+7	-3.3
Great Basin	14	5.5	5.0	0.1	1.8	6.9	+7	-1.4
Pacific North West	288	203.9	13.7	0.4	9.7	23.8	+264	+180.1
South Pacific	85	38.6	18.6	2.0	5.8	26.4	+59	+12.2

Deficit or strain situation with water in arid parts of both countries is superimposed by the trend of decrease of natural river run-off due to man's activity. From the point of view of the hydrological cycle, consumptive use of water occurs through evaporation, hence consumptive use increase means an increase in evaporation and decrease of run-off. Table 5 shows the corresponding data for the main rivers of the USSR.[3] The natural run-off of large Siberian and northern rivers has not changed and is not expected to change noticeably. As for the rivers in the south, however, the run-off decreased by 8% in 1970 and is expected to decrease another 30% by the year 2000. In some of the rivers, up to 80 or 90% of the water will be utilized.

Table 5. Decrease of river run-off in the USSR due to man's activity

River	Mean natural water resources		Decrease of annual run-off					
			1970		1981 - 1986		1991 - 2000	
	in the basin km^3/yr	at the mouth km^3/yr	km^3/yr	%	km^3/yr	%	km^3/yr	%
Volga	254	239	16	6	26	10	36	14
Ural	11.4	11.2	1.6	14	2.6	23	3.0	26
Terek	11.5	8.3	2.0	17	5.0	44	9.0	78
Kura	24.2	18.0	0.5	2	4.5	19	14.0	58
Don	27.9	27.9	5.1	18	8.5	30	12.6	45
Kuban	13.4	11.1	2.2	16	5.2	39	8.0	60
Dniester	9.3	9.3	1.3	14	3.0	32	3.5	38
Dnieper	53.5	53.5	10.7	20	13.8	26	18.0	34
Amudaria	77.1	41.0	5.0	12	24.0	59	39.0	95
Syrdaria	33.5	13.0	4.0	31	6.0	46	12.0	92
Ob	384	384	8.2	2	13.0	4	16.5	5
Zapadnaya Dvina	18.4	18.4	0.02	1	0.04	1	0.04	1
Severnaya Dvina	107	107	0.05	1	0.09	1	0.11	1
Pechora	128	128	0.002	0	0.005	0	0.007	0
Enisey	555	555	2.10	1	4.8	1	6.1	1
Lena	525	525	0.30	1	1.00	1	1.70	1
Amur	312	312	0.36	1	1.20	1	2.20	1
Total	2544	2462	54.5	2	118.9	5	183.1	7
For rivers of the Southern slope	516	432	42.4	8	97.6	19	155.1	30

Another water problem within the framework of IWT is the deterioration of lakes, internal seas, and estuaries.

As shown in Fig. 2, the main results of man's activity at a river level are:

(1) decrease of run-off due to increase of consumptive use of water;

(2) an increase in salt content and pollution due to agricultural and industrial activity;

(3) direct effect of reservoirs and dams.

At a lake/sea level it leads to increase of salt content and pollution, a decrease of lake level, and deterioration of conditions for aquatic ecosystems. These effects are typical and they are seen in the Great Lakes, San Francisco Bay, Gulf of California, Caspian Sea, Aral Sea, etc.

The decrease of inflow to the internal seas of the USSR is shown in Table 6.[3]

Due to the decrease of the natural inflow to the Azov Sea, the average salinity rose from 10–11 parts pro mille to over 12.5 parts pro mille;[4] this led to a considerable decrease of fish production, which indicates the need for fresh water to maintain the natural conditions of the ecosystem. The current fish catch of the Azov (6 tons/km^2) and Caspian Seas (5 tons/km^2) yields about 200 million roubles per year.[4]

Table 6. Decrease of surface water inflow to the internal seas of the USSR due to man's activity

Sea	Natural inflow	Decrease of inflow					
		1970		1981–1985		1991–2000	
	km^3/yr	km^3/yr	%	km^3/yr	%	km^3/yr	%
Caspian	295	22	8	44	15	74	25
Aral	54.0	9.0	17	30	56	51	94
Azov	41.1	7.7	19	14	34	21	51

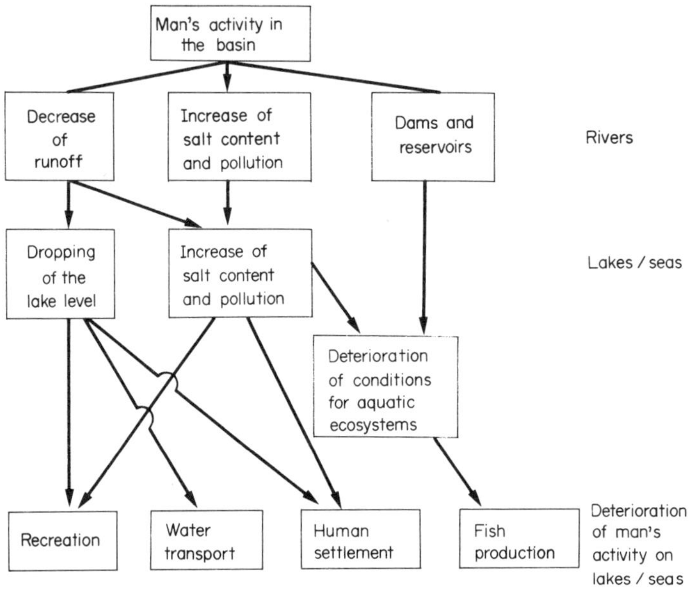

Fig. 2. Conceptual scheme of man's influence on lakes and internal seas.

The water level of the Caspian Sea dropped (due to natural causes) in the 1930s; this lead to a loss of about 15 or 18 thousand square kilometers of the fish productive area. The water level should increase now, unless a decrease of inflow due to man's activity develops. The water level is currently about −28.5 m a.s.l., and a drop of 0.8 or 1 should be regarded as catastrophic, as the area of highly productive shallow waters would diminish considerably and salinity in the northern part would change.[4] To improve the hydrologic regime, additional fresh water is required.

Therefore, between the arid regions of North America and Eurasia, there are a number of common features from the viewpoint of the demand/supply of water:

(1) a deficit of water exists in certain years and will continue to increase; moreover, the greatest consumptive use for water is irrigation;

(2) natural river run-off is declining due to man's activity; and

(3) a considerable amount of fresh water is required to maintain or improve ecological conditions of lakes and seas.

For the near future the southern slope of the USSR would require from 97 km^3/yr for a normal flow, to 120 km^3/yr for a low flow year (95% probability).[5] There are various strategies for the solution of this problem, and between them are situated IWT's. Interregional water transfers are appealing because of the great amount of water produced, which drastically changes the water situation. Large IWT's should be studied within a framework of strategic water management planning, at a continental or subcontinental level with the planning horizon not shorter than the years 2000 or 2030.

What is the problem structure of IWT? There are three main blocks: technology, socio-economy and environment. If there were three persons or three institutions responsible for a project and representing these three blocks, their objectives would all be different. The objective of the technology block would be to carry out the project and maintain its operation. The other two blocks would be constraints. The objective of the socio economy block would be to maximize a society's net benefit from a IWT; the benefit should not be expressed only in terms of money. The other two blocks would be constraints. The environment block's objective would be to ensure optimal conditions of the environment. As we do not usually know what these optimal conditions are, the objective would be to minimize disturbance in nature. The other two blocks would be constraints.

Thus, even at the highest level of the problem, a strong interrelation exists and the problem is of a systems nature.

The history of IWT projects reflects this structure. First, only technological engineering schemes were discussed and studied. Then, economic issues were incorporated. The next step should now be an integration of environmental issues and, therefore, the examination of the problem is in its present form.

Each block can be divided into sub-blocks, which are in turn also divided. Tentative division of the problem showing only the first and second levels is given in Fig. 3. Interrelations between blocks and sub-blocks are not shown and will be discussed later. Let us briefly discuss the sub-blocks in order to present some questions.

Technique – This means how to carry out and operate a new system.

Technology progress – Interregional water transfer projects would be designed to serve for approximately 100 yr. Hydrologic constructions of 1880 are now archaic. The rate of technology progress is increasing. The questions which then arise are: (a) How to take it into account? (b) Is it possible? and (c) Is it necessary?

Water use – Structure and amount of water use would change because of (a) IWT proper, and both its direct and indirect influence, (b) shifts in economy, (c) technology progress.

Resource consumption – These are manpower, capital investments, energy, materials, etc. to carry out the projects. It may be a very complicated strong subject for consideration, especially when it is necessary to choose between various large programs, only one of which would be the IWT.

Economic efficiency – This is obviously an important part of the overall problem and an important criterion in considering a project. Howe and Easter[2] successfully based their analysis of IWT efficiency on the following inequalities:

$$(DB_M + SB_M) + (DB_T + SB_T) > (DC_x + SC_x) + SC_c + TC,$$

$$TC + [(DC_x + SC_x) - (DB_T + SB_T)] < TC_A.$$

Let the benefits from the actual use of water be called direct benefits (DB), and the costs of giving up the direct use of water (i.e. benefits forgone) be called direct costs (DC). Benefits and

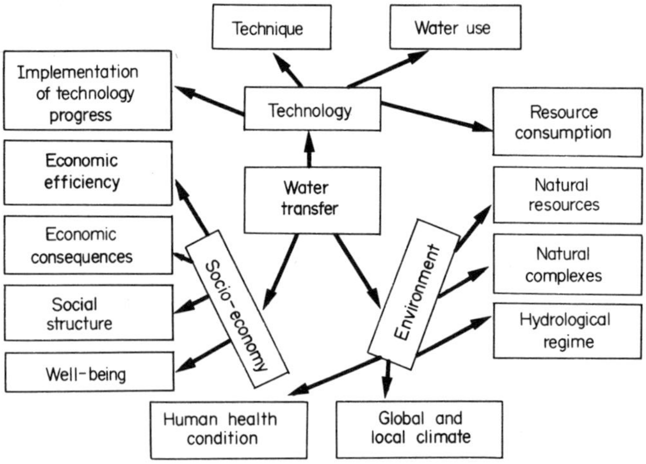

Fig. 3. Structure problem of water transfers.

> costs in market-related activities as seen from a national viewpoint will be referred to as secondary (SB, SC). TC represents the costs of the physical transfer system, and T_A the cost of the best alternative. Subscript *x*'s refer to parties in regions exporting water, *M*'s to those in regions importing water, *T*'s to affect parties in regions through which the transferred water will pass, and *c*'s to parties in regions whose outputs are competitive with those of the water-importing region (pp. 20–21).

And further,

> The first condition states that the increment to net incomes in the importing and transit regions must exceed the loss of incomes in the exporting region and in other regions where activities are displaced by the expansion of water-related activities in the importing region plus the costs of the physical transfer system, all properly capitalized on the basis of consistent time period. The second condition states that the cost of the physical water transfer scheme (including the net opportunity cost of the water) must be less than the cost of the best alternative for supplying the same amount of water to the importing regions. This comparison presupposes the prior optimum sizing of the projects (p. 21).

Is it, however, enough to measure the economic efficiency of a huge project just using this "conventional" technique? How does one evaluate long-term consequences? How does one take into account non-measurable values, including environmental values, and account for uncertainty?

Economic consequences – These are closely connected with matters of efficiency. The following questions arise: What are direct, secondary, and tertiary consequences, and what are they at a regional, national, and international level?

Social structure –– This would be greatly influenced by an IWT project both during its construction and operation. Many feed-back loops from the social structure and its change can be created, and influence various blocks of the problem.

Well-being – An overall objective of a water transfer project is to improve the people's well-being. How are we to understand the term of well-being? The meaning might be quite different.

Political and judicial issues of IWT are also quite important in a number of countries. However, this topic is mostly institutional and does not stem from the proper sense of the problem. For this reason, it has been omitted from the overall problem structure of IWT's. Practically speaking, the first steps of IWT projects are sometimes taken in these areas.

Natural resources – A large IWT project influences both renewable and non-renewable natural resources. Part of the resources can be used for construction (timber, mineral resources, etc.). Thereafter follow short-term impacts, such as the inundation of ore or oil fields, loss of areas with fertile soils, change of water quality due to new water reservoirs, etc. These are followed by long-term consequences, i.e. biological changes which include forest productivity, etc. The question is: how then, to assess an influence of projects on natural resources?

Hydrologic regime, climate and natural complexes – These are blocks describing the corresponding processes in the hydrosphere and atmosphere, and in the interface between the hydrosphere, atmosphere, lithosphere and biosphere. These blocks are another face of "Natural Resources", the latter being in many cases an economic aspect of the former. There are many questions to consider. An indispensable part of a IWT scheme is a water reservoir. Its creation means losses of the best lands, deterioration of water quality, etc. Reallocation of water means change of water circulation in the atmosphere, that is, a change of climate – but to what extent? A natural complex (geoecosystem) consists of hundreds and thousands of interrelated elements. A change in one of its components or groups would lead to many other changes in a geoecosystem. Assessing the project, it is very important in evaluating environmental natural features. This brings up the question: how to measure them and how to compare non-commensurable phenomena. We must also determine a way to predict environmental changes in both quick impacts and long-term cumulative responses, and those which may be disregarded.

Human health –– It is well known that human health is a product of both natural and social factors. Changes in both groups of factors would lead to impacts in human health.

Sub-blocks discussed here can be disaggregated. There are many relations between partial issues concerning the IWT problem. The whole problem can be represented as a trihedral pyramid, each side of which is one of the main blocks (technology, socio-economy, environment). Each block ramifies according to the level of aggregation of the problem, and these ramifications have many links both within the same main block, as well as with other blocks (sides). Hence, the problem is really interdisciplinary.

One of the authors has tried to single out the most important relations between the sub-blocks, and to evaluate the intensity of them. The results are shown in the matrix (Table 7).

Two principal conclusions can be drawn from this matrix: (1) there is a strong interrelationship within each of the main three blocks; (2) the relationship between the blocks is no less strong. Thus, to use an expert evaluation approach as a tool to develop and assess an IWT project, it is necessary to organize experts' work not only within the same field, but to implement from the very beginning an interlinkage between various fields.

The problem of assessment of large IWT projects is very complicated – we do not have examples of developed methodology. Another question arises and may be discussed during our meeting, but can hardly be answered: is there a developed methodology for other large projects? A certain amount of experience was gained, for example, during elaboration of the Alaskan Pipeline Project. By the term methodology, we understand here an elaborated system of techniques applied to each IWT project rather than a loose set of approaches. Other questions to put forward for discussion are: (a) What method of assessment can be used? (expert evaluation? simulation games? optimization models? simulation models? others?);

Table 7. Interactions between various factors in assessment of water transfer projects

Action		Impact	Technology				Socio-economy				Environment				
			1	2	3	4	5	6	7	8	9	10	11	12	13
Technology	1	Transfer technique			+ +	+ + +	+ + +				+	+	+ + +	+ + +	+ +
	2	Technology progress	+		+ +		+ +	+							
	3	Water use	+	+			+ + +	+ +	+		+		+ +	+ + +	+
	4	Resource consumption	+ +	+			+ + +	+ +		+					+ +
Socio-economy	5	Economic efficiency	+ + +	+ +	+ + +	+ + +		+ +	+ +	+ + +				+ +	+ +
	6	Economic consequences		+	+ +	+	+ +		+ +	+ +	+				+
	7	Social structure			+ +	+									
	8	Well-being	+ +	+	+ +	+	+		+ +		+ +			+	+
Economy	9	Human health	+		+ +					+ +					
	10	Climate									+		+ + +	+ +	+
	11	Hydrologic regime	+ +		+		+ +				+			+ +	+
	12	Geoecosystems					+ +						+ +		+ +
	13	Natural resources	+ +			+ +	+ +	+					+	+ + +	

(b) Do steps in a project's development correspond to a certain method? (c) What is an approach to integrate various disciplines in an assessment? (d) How is it organized?

The list of questions can be expanded. After all, there are two goals of this meeting: (1) to discuss our knowledge of the IWT problem; and (2) to clarify questions for further studies of this very complicated problem.

REFERENCES

1. Zarubaev, N.V. *Multipurpose Use and Conservation of Water Resources,* Stroiizdat, Leningrad, (in Russian), (1976).
2. Howe, C.W. and Easter, K.W. *Interbasin Transfers of Water: Economic Issues and Impacts,* Johns Hopkins Press, Baltimore, MD, (1971).
3. Voskrensensky, K.P. *et al.* Man's Influence on Water Resources and Hydrologic Processes, *Proceedings of the IV All-Union Hydrological Congress,* Vol. 1, Leningrad, (in Russian).
4. Marti, J.J. and Ratkovich, D.Y. Water management problems of Azov and Caspian Seas, *Vodn. Resur.,* 3, (in Russian), (1976).
5. Gerardi, I.A. *State* Systems of Regulation and Interbasin Reallocation of Water Resources of the USSR, *Proceedings of the IV All-Union Hydrological Congress,* Vol. 4, Leningrad, (in Russian), (1976).

PART I: Survey of Proposed Projects on Interregional Water Transfers

Water Supply & Management, Vol. 2, pp.79-90. 0364-7714/78/0601-0079$02.00/0

North American Water Transfers: An Overview

ASIT K. BISWAS*

INTRODUCTION

Interregional water transfer is not new: it has been practised from time immemorial. For example, the ancient Egyptians diverted river water over long distances several thousand years ago.[1] But its importance has increased in recent years, especially as population pressures in many arid regions of the world have made it imperative to grow more food. Agricultural production can be increased in two ways – by increasing crop yields and by bringing new land under cultivation. Both of these alternatives can only be viable, provided adequate water supply is available.

The importance of water control for crop production can be illustrated by the following facts. On a global basis, agriculture uses 80% of all water consumption: the corresponding figure for the United States is about 40%. Total irrigated area in the world is 233 million ha, out of which 93 million ha are in developing countries.[2] Irrigated land constitutes less than 10% of global cropped area, and yet it accounts for 30–40% of total agricultural production. Thus, as populations in developing countries continue to increase, and since these countries are without exception in the tropics and subtropics, water control is increasingly becoming a major requirement to boost food production. As water resources of populated regions become fully developed, interregional transfer becomes an attractive possibility – provided the environmental and social problems associated with such major projects can be resolved.

Interregional transfer is, however, one of several alternatives of non-conventional water development. There are other possibilities, among which are weather modification, desalination, iceberg towing, and the use of VLCC (very large crude carriers) to transport fresh water to water-deficient regions. None of these are universal solutions, and each must be considered in relation to problems of the region being analysed. In other words, these solutions are site-specific. Furthermore, for most of these unconventional techniques, there exist technological, legal and environmental problems many of which have yet to be solved. In many cases, economic constraints have yet to be overcome.

INTERREGIONAL WATER TRANSFER

First, what is meant by water transfer? On simplest terms, it may be defined as inter-basin diversion, that is, the artificial withdrawal of water by ditch, canal or pipeline from its source in one contributing or exporting drainage basin for use in another receiving or importing basin. Before such interregional diversion can be seriously considered, it is important that

* Director, Biswas & Associates, 3 Valley View Road, Ottawa, Ontario, Canada, K2H 5Y6, and International Institute for Applied Systems Analysis, Laxenburg, Austria.

three factors be analysed. These are assessments of available water resources, the nature of the demand function and the efficiency of water use.

Assessment of available water resources is an important factor for any water resources development plans. Availability of water varies with space and time. Thus, before reliable forecasts of water availability can be made, it is necessary to have adequate data over a reasonable period of time. Based on such data, long-term development and management plans can be established. However, in many parts of the world, such data are not available, or if available, they are for a rather limited time horizon. This is unfortunately the case for most developing countries. Even for North America, adequate data are not always available – especially for northern Canada and many parts of Mexico. Without a comprehensive assessment of water availability, it is difficult to contemplate interregional water transfer – or any other water planning process for that matter.

Second important consideration is the assessment of water demands for various purposes. The term "demand", in the context of water management, really means requirements, and is very rarely used in its traditional economic sense. Indeed, very rarely is the concept of demand elasticity explicitly considered within the water planning process. Thus, it should be no surprise that very little is known about constructing appropriate demand functions under varying socio-economic considerations. In other words, emphasis so far has been on supply management – that is, increase in supply is considered to be virtually the only management alternative – rather than on demand management. As the demands for water for various purposes continue to increase and available sources become more and more exploited and polluted, emphasis will have to gradually shift from supply to demand management.

Efficiency of existing water use is the third important consideration. Agricultural sector is an inefficient user of water, and where most improvements could be made. Existing efficiencies of irrigation systems are so low that they do not by any means reflect the actual water requirements of crops. On a global basis, 1.3 million million m^3 of water is used for irrigating crops, but for this 3 million million m^3 of water have to be withdrawn.[3] In other words, 57% of water withdrawn is lost in the distribution system. This, however, does not mean that 43% of water reaching irrigated fields is efficiently used. Over-irrigation is not exactly an uncommon practice in both developed and developing countries. It not only means that water is wastefully used but also contributes to development of adverse environmental problems, like increase in groundwater table and salinity levels.[4] Thus, before major alternatives like interbasin transfer can be analysed, the possibility of optimization of water use by increasing efficiency should be considered. As a rule, it is cheaper to obtain more water per unit cost by improving the efficiency of water use from existing projects than from building new ones. Also, the time required to plan and to build new schemes is significantly longer: the efficiency of existing projects can be improved more quickly.

INTERREGIONAL TRANSFERS IN NORTH AMERICA

To bridge the gap between past images of development and proposals for new construction, it is important to recognize the groundwork of water diversions already established between river basins. Figure 1 illustrates this pattern aggregated for major basins in the United States. The figure shows that the walls of the Colorado basin have obviously been breached in a number of places for exporting water to surrounding regions, but the Columbia and the North Pacific, on the other hand, have remained largely self-contained and water-abundant.

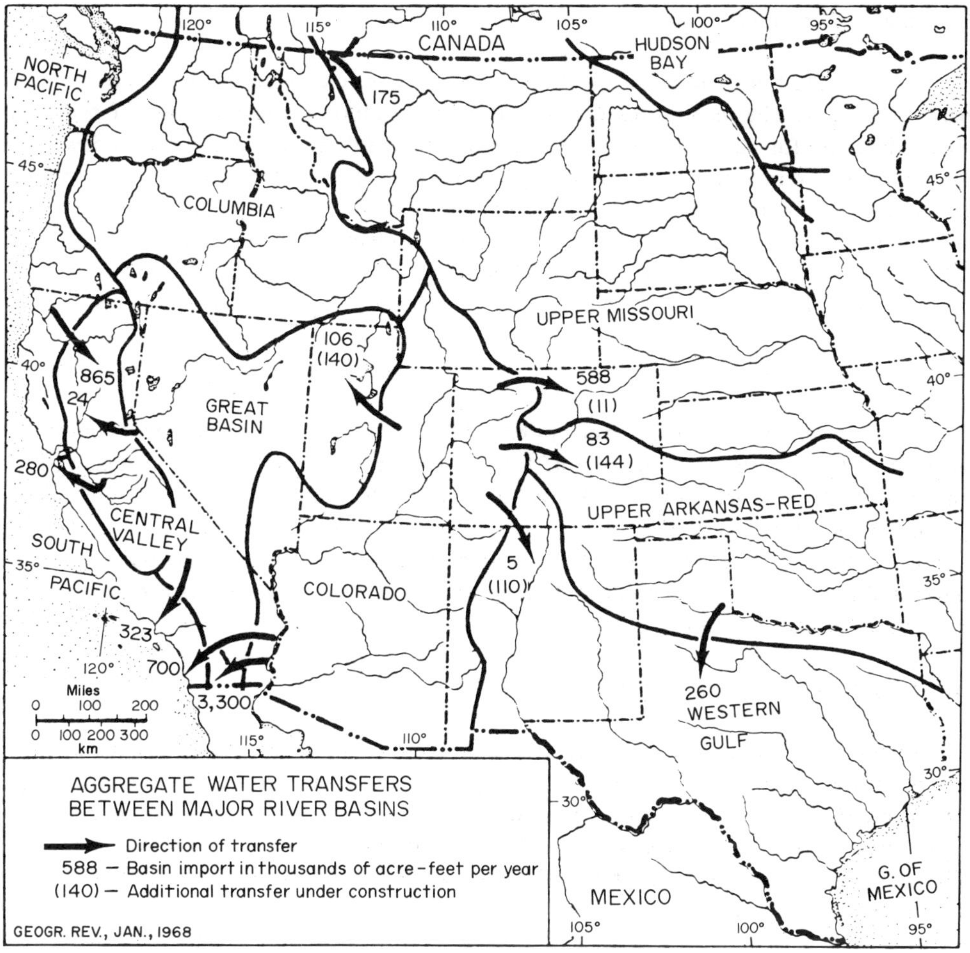

Fig. 1.

Late-developing urban centers, unable to dislodge agricultural water rights, account for most of the recent long-distance water importations. Probably one person in three in the Western United States is now served by a system which imports water from 100 or more miles away. It is quite likely that Los Angeles, Salt Lake City, Laramie, Denver and Colorado Springs would have found it impossible to each beyond their own river systems to the Colorado, if the diversions had meant crossing their state lines as well.

The same can be said for the rest of the continent. New York City takes from the Delaware; Boston from the Connecticut; Chicago from Lake Michigan into the Illinois; the province of Ontario from the Albany and Kenogami drainage of Hudson Bay into Lake Superior; British Columbia from the Nechako to the Coast at Kitimat. Their effects, of course, have sometimes been felt downstream across the boundaries.

A few select interbasin diversions will be briefly discussed herein. These have been selected primarily on the basis of their magnitude or historical importance. The experience in California or Mexico will not be discussed, since they are discussed elsewhere in this collection.

1. *Nechako–Kemano Diversion*

In 1925, the Kenny Dam on the Nechako River was completed by the Aluminum Company of Canada. The dam stores the run-off from a 5400 square mile area of the Fraser River drainage for diversion westward to the company's power plant near tidewater at Kemano. The average flow, approximately 6500 cfs at the point of diversion, is controlled by the 6.6 million acre-ft of usable storage in the reservoir. Diversion is made by means of a 10.1 mile long, 25 ft diameter tunnel leading to an underground powerhouse with a head of 2500 ft and a present installation of 707,000 kW. The average annual diversion from the reservoir is about 3300 cfs. Excess water is spilled or released to the Nechako–Fraser system.

2. *Lake St Joseph Diversion*

Flows from the Winnipeg River system of Western Ontario and Eastern Manitoba are augmented by a diversion of the run-off from a 4760 square mile area of the headwaters of the Albany River which flows north east to Hudson Bay. Water is diverted from Lake St Joseph in the Albany system southward to Lac Seul and onward down the English River and Winnipeg River to Lake Winnipeg. The diversion increases the offpeak flows on the Winnipeg River and raises power production at nine hydro plants. The diversion has averaged approximately, 2800 cfs since 1957.

3. *Long Lake and Ogoki Diversions*

In addition to the Lake St Joseph diversion, there are two other diversions of water from the headwaters of the Albany River. The Long Lake and Ogoki diversions came into being in 1939 and 1943, respectively, and re-routed Albany River water southward to Lake Superior, and hence assists Great Lakes water level control during low water years. The Long Lake diversion has averaged approximately 1400 cfs and the Ogoki diversion about 3900 cfs which taps a basin of 5800 square miles.

4. *Chicago Diversion*

Diversions of water from the Great Lakes–St Lawrence system at Chicago into the Mississippi River system have been made since 1848. Diversions reached magnitudes of slightly over 10,000 cfs in 1928 but the diversion at present is governed by United States Supreme Court decree of 12 April 1930, which provided that on and after 31 December 1938, the diversion would be limited to 1500 cfs in addition to domestic needs of the City of Chicago. The present diversion averages 3200 cfs. A new International Joint Commission (IJC) reference has been announced, authorizing a 5-year study of the effects of increasing the diversion up to a maximum of 10,000 cfs for abstraction of the water of Lake Michigan at Chicago, which will have an effect on the water levels and flow of the Great Lakes.

5. *Chamberlain Lake Diversion*

While the Chamberlain Lake diversion does not involve waters of Canadian origin it warrants comment because of its very interesting history and the role it played in pointing out the need for a joint United States–Canada commission to deal with international rivers.[5]

6. *Garrison Diversion*

The Garrison diversion is part of the Pick–Sloan Plan, a huge development plan proposal for the Missouri River Basin. The immediate purpose of the Garrison project, which is under construction, is to direct 879,000 acre-ft of water annually from the Missouri River for municipal, industrial and recreational use, and for irrigation of 250,000 acres of land. The irrigation aspect is the principle cause for concern due to the possible introduction of foreign species of fish and biota to Canadian waters, as well as wastes. Most of the land to be irrigated is on the northern side of the continental divide and drains into the northward flowing Red and Souris Rivers.

7. *Saskatchewan–Nelson Basin Study of Possible Diversions*

In 1972, the Saskatchewan–Nelson Basin Board completed a 4-year study of Canadian Prairie rivers in terms of additional supply of water by diversion or storage in one of the four largest river basins in North America (414,000 square miles). The Basin includes three major river systems, the Saskatchewan, the Red and the Winnipeg. Preliminary engineering reports identified 55 possible dams and 23 diversion projects. The design of a water demand study for the Basin has been completed and will be implemented in the near future.

8. *James Bay, Quebec*

This diversion is centered on the La Grande River, and reflects drainage basins totalling over 64,000 square miles made up of La Grande (37,800 square miles) plus parts of three other rivers: the Eastmain (by diversion of the Opinaca, a tributary), the Great Whale (flowing into southern Hudson Bay), and the Kaniapiskau–Koksoak (flowing into Ungava Bay). The first phase of the La Grande scheme, which is currently under construction, will produce in excess of 8000 MW (12,000 MW ultimate) and calls for an investment in excess of 6 billion Canadian dollars.

9. *Lake Winnipeg, Churchill and Nelson Rivers*

A two-year $2 million study was completed in 1974 to determine the effects that regulation of Lake Winnipeg, diversion from the Churchill River and development of hydro-electric potential of the Churchill River would have on water and other resources. The diversion of up to 30,000 cfs from the Churchill River has currently reached 20,000 cfs. Six out of the fourteen hydro-electric stations have been constructed at Jennep producing 168,000 kW of power. The diversion will reduce the Churchill River discharge from 40,000 cfs to 13,000 cfs. An implementation agreement covering all of the study recommendations is currently under negotiation by the Canadian Government and the Province of Manitoba.

10. *Dickey–Lincoln School Project*

This hydro-electric project has been in the planning stages since the early 1960s.[6] The project consists of two dams with generating facilities in each. A large dam at Dickey (Maine), some 15 miles upsteam of the international reach of the St John River (at the International Maine/New Brunswick border), would create a reservoir 88,000 acres in area. The total volume of the proposed reservoir is about 8 million acre-ft, of which 2.9 million acre-ft are live storage. The powerhouse at Dickey would generate 760 MW for peaking purposes, while the second dam at Lincoln School, 1.5 miles upstream of the international reach, would be used to even out fluctuations in outflow from Dickey. About 5000 acres would be flooded on tributaries in the Province of Quebec. An environmental impact study costing $750,000 is currently underway for this $500 million project.

11. *Colorado River Basin*

In 1968, as part of the Colorado River Basin Project Act, Congress prohibited federal studies of importation of water to the southwestern United States for a period of 10 years. That moratorium expires in 1978. Southwest water interest groups continue to seek studies of alternatives for importation of water from the northwest. Added pressure due to increasing droughts will undoubtedly occur when the moratorium expires.

PROPOSALS FOR INTERREGIONAL TRANSFERS

Proposals for interregional transfers have been made in the 1960s, the most ambitious of which is the North American Water and Power Alliance (NAWAPA), first proposed in 1964. The immensity of the plan stirred the imagination of many engineers and economists, and within the 5 years of NAWAPA being proposed, a whole series of interregional water transfer schemes were put forward to redistribute the waters of North America. Table 1 shows twelve such schemes which are primarily national in character. Eight major international proposals are shown in Table 2.[8] Only the major one, NAWAPA will be briefly discussed.

NAWAPA project – This $100 billion project was proposed by Ralph M. Parsons Company of Los Angeles.[9] The general idea is to collect surplus water from the high precipitation areas of the northwestern part of the North American continent and distribute it to water-scarce areas of Canada, the United States and northern Mexico. A series of dams and power stations in Alaska and northern British Columbia would collect water and provide power to pump this water up to the Rocky Mountain Trench Reservoir in southeastern British Columbia. From the Rocky Mountain Trench Reservoir, water would be lifted by pumps to the Sawtooth Reservoir on Central Idaho. From there, the water would flow by gravity to the western States. NAWAPA would initially provide 137.5 billion m^3 of water annually to seven provinces of Canada, 33 states in the United States and three northern states of Mexico. The total power generation would be 100 million kW/yr. Out of this, 30 million kW/yr would be utilized by the pumping requirements of the project. NAWAPA is a gigantic project and its environmental and social costs have yet to be determined. In the present era of environmental awareness, it is highly unlikely that such a major project will receive serious planning attention – at least for another two decades.

ISSUES INVOLVED WITH INTERREGIONAL TRANSFERS

There are a number of problems associated with (and several issues involved in) large scale transfers of water. These are discussed below under four separate headings: technical, socio-economic, political and legal, and environmental.

Table 1. Interregional transfer proposals (national)

Proposal (Author)	Year proposed	Water source	Volume of diversion in millions of acre-ft	Estimated cost in billions of $
Pacific Southwestern Water Plan (Interior Dept)	1963	north coastal California	1.2	?
Western Water Project (Pirkey)	1963	lower Columbia at Dalles	13.0	12.8
Snake - Colorado Project (Nelson)	1963	middle Snake in Idaho	2.4	1.4
Modified Snake - Colorado Project (Dunn)	1964	lower Snake in Oregon	5.0	3.6
Yellowstone - Snake - Green Project (Stetson)	1964	Yellowstone and Snake, Montana and Idaho	2.0	0.4
Undersea Coastal Aqueduct (NESCO)	1965	Klamath, Eel and Rogue, mouths	11.0	8.0
Texas Water Plan (State of Texas)	1965	eastern Texas rivers	3.3	0.5
Prime Plan (Province of Alberta)	1965	Peace and Athabaska rivers	?	?
Mexican Plan (Government of Mexico)	1965	southern east coastal region	?	?
Undersea Hose (Conner)	1967	mouth of Columbia	12.0	2.0
Beck Plan (Beck)	1967 - 1968	Missouri in Nebraska	10.0	3.5
Hudson Institute Plan (Hudson Institute)	1968	Mississippi and Arkansas	34.0	12.2

Table 2. Interregional transfer proposals (international)

Proposal (Author)	Year proposed	Water source	Volume of diversion in millions of acre-ft	Estimated cost in billions of $
Grand Canal Plan (Kierans)	1959	James Bay dyked rivers "recycled" to Great Lakes	?	?
Great Lakes - Pacific Waterways Plan (Decker)	1963	Skeena, Nechako and Fraser of B.C., Peace, Athabaska, Saskatchewan of Prairie Provinces	115.0	?
North America Water and Power Alliance, NAWAPA (Parsons)	1964	Primarily the Pacific and Arctic drainage of Alaska, Yukon and B.C.; also tributaries of James Bay	110.0 initially	100
Magnum Plan (Magnusson)	1965	Peace, Athabaska and North Saskatchewan in Alberta	25.0 at border	?
Kuiper Plan (Kuiper)	1967	Peace, Athabaska and North Saskatchewan in Alberta Nelson and Churchill in Manitoba	150.0	50
Central North American Water Project of CeNAWP (Tinney)	1967	Mackenzie, Peace, Athabaska, N. Saskatchewan, Nelson and Churchill	150.0	30 - 50
Western States Water Augmentation Concept (Smith)	1963	Primarily Liard and Mackenzie drainages	38.0 at border	75
NAWAPA + MUSHEC or Mexican - States Hydroelectric Commission	1968	NAWAPA sources + lower Mississippi and Sierra Madre, Oriental rivers of southern Mexico	158 + 129 NAWAPA MUSHEC	?

Technical

(i) The planning, design and construction of gigantic projects for transporting large bodies of water over large distances must be carried out with great care and imagination by highly qualified individuals. This is necessary because as a rule the bigger the size of such projects, the greater are the uncertainties. It may be more prudent to postpone a decision on such projects, if all investigations are not complete, or if uncertainties are enormous – rather than make decisions in a hurry. Also, such large projects can be severely affected by upstream and downstream developments, i.e. sudden failure of a hydraulic structure upstream can adversely affect the overall safety of the project.

(ii) The hydrologic and meteorologic characteristics of the drainage basins may significantly change after the completion of the project. Such changes should be adequately anticipated and considered within the planning framework.

(iii) For determining surplus water of a basin, its long-term storage requirements to take care of the time variation of run-off in the basin should be considered. Regions with a present water surplus have been understandably reluctant to permit exports which might have even a slight probability of restricting their own economic growth in the distant future. There have been controversies over diversions of the Colorado,[9] Columbia[10] and Yukon[11] rivers in the United States. However, if the population of such regions can be assured that only surplus water of their basins, calculated after taking into account their projected requirements (say, for a period of 50 years as provided in the Texas Water Plan),[12] as well as their storage requirements for taking care of the time variation of the run-off at a given location in the basin, will be diverted, the objection could perhaps be minimized.

(iv) The project should be made flexible by leaving as many options as possible open for future adjustments decisions. Large-scale transfer of water from one basin to another is, in effect, an interference with the natural water regime. For such cases, from ecological and environmental viewpoints, a cautious and conservative approach is desirable. In spite of scientific and technological developments of recent years, not much is known about the behaviour of the streams and rivers under changing flow and sediment conditions. Thus, a number of uncertainties are involved in the planning and design of such large-scale projects. It may, therefore, be desirable to make decisions on the project in various stages and the plan be kept flexible enough that only those decisions which are essential for the immediate future have to be made.

(v) Efforts should be made to determine the impact of interbasin transfer of large bodies of water on the environmental characteristics of the region. When surplus water of a river is diverted, the waste assimilative capacity of the river decreases. This may adversely affect the biological life in the downstream reaches.

(vi) Interbasin transfer routes and reservoir sitings must take into consideration earthquake-prone areas. Safety should be a significant concern of such projects.

Socio-economic

(i) Howe and Easter[13] have concluded that large-scale transfers of water are likely to cost more than they are worth to a nation, except in certain "rescue operation" cases where diverting water supplies threaten to idle immobile capital and labour. On the other hand, Wells[14] points out that the importation of water to the high plains of Texas is not economically feasible but also that the State simply cannot afford not to import water to the area. Thus, there can be great diversity of results that can be obtained from the economic analysis of such large projects. Sufficient care and effort must therefore be exercised in economic feasibility studies.

(ii) A large-scale interbasin transfer project must be justified not only in terms of the direct costs of transporting water, but also in terms of the value of services foregone by the exporting region due to the diminuition of its water supply (the opportunity cost of the diverted water).[15] Various alternatives to interbasin transfers should be investigated and that alternative which provides water to the deficient areas at the minimum cost (including special and environmental), should be selected. Alternatives might include:

(1) more efficient use of water within existing allocation patterns;
(2) reallocation of surface supplies;
(3) wastewater reduction, including desalination;
(4) improved integration of surface and groundwater supplies;
(5) management of watersheds; and,
(6) weather modification.

(iii) As an alternative to importation of water, economy in the water use should also be considered by the importing region. The philosophy of controlling water demands to eliminate or reduce further water development involves a number of factors and has been discussed by Coe[16] in detail.

(iv) The framework for the economic analysis should be properly made; not only the primary benefits, but also where possible, the secondary and tertiary benefits of the projects should be considered.

(v) Broad social objectives and benefits should be considered in the light of growing urban and industrial demands; while benefits to the agriculture sector are usually the major concern of the majority of diversion schemes, social objectives such as income redistribution, alteration in regional growth rates, reduction in unemployment and environmental protection (though not always measurable in purely economic terms) should also be considered in assessing alternative public investments *in situ* demands e.g. fish, aquatic vegetation, power, recreation and navigation should also be considered.

Political and legal

(i) In many Western States, the earliest water rights were developed simply by use and some are still unrecorded. Even in the areas where all rights have been adjudicated they are measured in quantity of water withdrawn, rather than in quantity actually consumed, which must be determined before a transfer can take place.[17]

(ii) It is more than coincidence that all transfers of water that have been effected thus far on the continent, fall within state and provincial borders. While the effects are certainly felt downstream across these borders, it is still fair to say that the present pattern of interbasin diversions strongly reflects the potential regionalization of Canada and the United States.

(iii) The foundation of Canadian–American international water law derived from the 1895 United States–Mexico dispute over the use of the Rio Grande. The United States reply is what is now referred to as the Harmon Doctrine.[18] A less stringent version of the same doctrine or principle appeared in Article II of the 1909 Boundary Waters Treaty.

(iv) Of all the difficulties associated with major interbasin transfers, the legal and political considerations are the most complex.

(v) The Supreme Court decision of *Arizona* vs *California* (373 US 566, 1963) was a landmark decision in federal–state water law. During the 10 years of litigation (1952–1963) both parties came to the conclusion that the water supply of the Colorado River was not big enough to supply the needs of both states, and therefore a supplemental source of water had to be found outside the Colorado basin.

Environmental

(i) The environmental implications of large interregional water transfer schemes are many, and these should be carefully analysed and received. A major difficulty is to assign economic values to many of the social and environmental costs stemming from such projects.[19] These should at least be subjectively evaluated.

(ii) The increase in both water-borne and water-based diseases due to major water developments, especially in the tropics and subtropics, should be carefully considered. Appropriate countermeasures should be taken to ensure that incidence of such diseases are kept to a minimum.[20]

(iii) Major interregional transfer schemes may affect the flora and fauna, and may cause irreparable damages.

(iv) Changes in micro- and macro-climate due to large-scale developments is always a possibility, especially in terms of increased evaporation and fog formation.

(v) Since the promulgation of the US National Environmental Policy Act (NEPA) of 1969 and the issuance of the Canadian Environmental Assessment Guidelines (1976), it is necessary to prepare environmental impact assessment for all significant federally funded projects. This includes, *inter alia*:

the environmental impact of the proposed action;
any adverse environmental effects which cannot be avoided should the proposal be implemented;
alternatives to the proposed action;
the relationship between local short-term uses of man's environment and the maintenance and enhancement of long-term productivity; and
any irreversible and irretrievable commitments of resources which would be involved in the proposed action should it be implemented.

CONCLUSIONS

Within the North American context, the following conclusions can be drawn for future large-scale interregional water transfer projects.

(1) Opposition to water exports, especially for interstate and international projects is likely to increase, especially as water becomes a scarce commodity. Logically this is hard to explain, since states and countries freely export other resources like minerals, hydrocarbons or agricultural products. In fact, for these resources, emphasis seems to be on increasing exports. Public sentiments, for some reasons, seem to be against water exports, and this is reflected within the political process. This is unlikely to change in the near future.

If the plan is self-contained within a state, its probability success is much higher.

(2) There is a tendency within the engineering and economic professions to opt for technological solutions – "soft" options are seldom seriously considered. Since water

resources development is dominated by these two professions, there is a tendency to make decisions to go ahead with technological fixes before all the alternatives are explored.

(3) The legal aspects of interstate and international transfers are quite complicated. This can be easily noticed when the number of serious disputes arising out of management of interstate and international rivers and lakes are considered. This is a global, and not exclusively North American, problem.

(4) Since the late 1960s opposition to major interregional water transfer projects has increased on environmental grounds. In the United States, it is unlikely that such schemes will be implemented within the forseeable future – at least for the next two decades.

REFERENCES

1. Biswas, Asit K. *History of Hydrology*, North-Holland, Amsterdam, (1970).
2. Biswas, Asit K. Editor's Introduction in *United Nations Water Conference: Summary and Main Documents*, Pergamon Press, Oxford (1978).
3. Biswas, Asit K. Water and man, *Water Int. J. Int. Wat. Resour. Ass.* (1978).
4. Biswas, Asit K. Environmental implications of water development for developing countries. *Water Supply& Management*, **2** (4), (1978).
5. Clark, R.H. *Canada's Surface Water Resources*, Reprint Series No. 50, Dept. of Environment and Fisheries, Ottawa, Canada, (1967).
6. *The International Passamaquoddy Tidal Power Project and Upper St John River Hydro-electric Power Development*, US Army Corps of Engineers, (1964). See also *United States–Canada Joint Study of the St John River*, St John River Engineering Committee Preliminary Report on Stage 1, *Storage Benefits*, St John River Engineering Committee, (1965).
7. The Ralph M. Parsons Company *NAWAPA – North American Water and Power Alliance*, Brochure No. 606–2934–19, Los Angeles, California, (1965).
8. Quinn, Frank J. *Images of Continental Water Engineering, The Continental Use of Arctic-flowing Rivers*, Tinney, E.R. (ed), Water Research Center, University of Washington, Seattle, Washington, (1968), 176 pp.
9. White, Gilbert *et al. Water and Choice in the Colorado Basin*, Publication 1689, National Academy of Sciences, Washington, DC, (1968), 107 pp.
10. Krutilla, J.V. *The Columbia River Treaty: The Economics of an International River Basin Development*, Johns Hopkins Press, Baltimore, MD, (1968).
11. Schram, Gunter. The economics of an Upper Yukon basin power development scheme, *Ann. Regional Sci.* **2** (2), (December 1968).
12. Texas Water Development Board *The Texas Water Plan*, Austin, Texas, (1968).
13. Howe, C.W. and Easter, K.W. *Interbasin Transfers of Water: Economic Issues and Impacts*, Johns Hopkins Press, Baltimore, MD, (1971).
14. Wells, D.M. High plains irrigation and Texas water plan. *Am. Soc. civ. Engrs., J. Irrig. Drain Div.*, pp. 123–130, (March 1971).
15. Moncur, J.E.T. Opportunity costs of a transbasin diversion of water: methodology. *Water Resour. Res.* **8** (6), 1415–1422 (December 1972).
16. Coe, J.J. Satisfying water needs in Southern California, *Am. Soc. civ. Engrs, J. Irrig. Drain Div.*, 483–499, (September 1971).
17. Ellis, Willis H. Water transfer problems: law in water, in *Water Research*, Kneese, A.V. and Smith, C. (eds) Johns Hopkins Press, Baltimore, MD, (1968), pp. 233–248.
18. Austin, Jacob *Canada–U.S. Practice*, Canadian Bar Review, B–93, (1959), p. 408.
19. Biswas, Asit K. and Coomber, N.H. *Evaluation of Environmental Intangibles*, Genera Press, Bronxville, NY, (1973), 77 pp.
20. Biswas, Margaret R. Environment and food production, in *Food, Climate and Man*, Biswas, M.R. and A.K. (eds), John Wiley, New York, (1978).

Water Supply & Management, Vol. 2, pp.91-101. .0364-7714/78/0601-0091$02.00/0

The Scientific Principles of Large-scale Areal Redistribution of Water Resources in the USSR

G. V. VOROPAEV*

The high rates of productive forces development in the Soviet Union have given rise to a large increase in water consumption of all branches of the national economy. Only for the past 5-year period, 1970–1975, the total withdrawal from water sources has increased by 20%. It is apparent that in the near future, water consumption rates will persist and by the end of the century they may amount to 600–700 km^3 year, exceeding the present value by 2–2.5 times. Appreciable changes in both the hydrologic regime and water quality occur because of the increase in water withdrawal and return water diversion and related construction of hydraulic structures and man's activity on watersheds. It affects adversely the ecological systems of water bodies and ecosystems at places of water use. A number of regional water problems have emerged and get increasingly complicated, and these problems will become world-wide in the course of subsequent water resources development.

The uneven distribution of river run-off, which is the main source of water resources, makes the solution of new water problems in our country exceedingly difficult. The southern areas, with up to 63–75% of the country's water consumption, have less than 15% of run-off. Adjacent and inland seas are another complicated factor: it places stringent requirements on the amount and quality of water.

Among the solutions of the above problems are: (1) all-round decrease in water consumption by improving water use technology in all branches of national economy and replacing a number of water consuming industries by waterless ones; (2) more complete use of all local water sources and enlargement of their resources by watershed management; (3) interregional water transfer by diverting part of the run-off of northern rivers to the basins of southern rivers.

The present state and perspective estimation of the scientific and technological progress in water consumption and water resources management indicate the possibility of achieving certain practical results in the way of reducing specific indices of water consumption, increasing the degree of water treatment and reuse, replacing water-demanding industries, etc. However, in many regions of the USSR, water requirements will exceed water availability. Water deficit in sourthern areas may be over 100 km^3/yr by the end of the century. To cope with these requirements in such a short period of time is possible only by undertaking a complex of measures including interregional water transfers.

The scale of construction needed for the interregional transfer of water resources in the USSR to meet the demands of the national economy at the end of the century and later may be so enormous and the impact on natural processes in major regions may be so strong and

* Institute of Water Problems, USSR Academy of Sciences, Moscow, USSR.

irreversible that now it is impossible to foresee the effect of the interregional water transfer on the environment and predict ecological and social consequences without special studies. Despite the existing valuable experience in solving large-scale water resources problems in the Soviet Union, we have no acceptable criteria for solving the newly arising problems and selecting the best project alternatives. The economic criteria established by practice and being used at present proved to be insufficient.

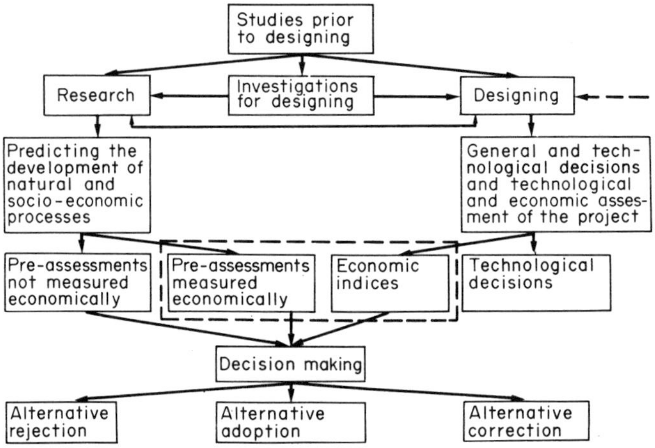

Fig. 1. The interrelationship between designing and research in solving a large-scale water resources problem.

Figure 1 shows the inter relationship between designing and research in solving a large-scale water resources problem. Interregional water transfer is such a problem. The experience of design and hydraulic structure construction convinces us that the larger the project, the more complicated are its relations with the environment and economy, and thus technological and economic assessment alone prove to be insufficient for making decisions. Therefore, special studies for predicting the development of natural and socio-economic processes, caused by water resources development, are necessary. Such studies will permit obtaining both qualitative and quantitative pre-assessments of these processes. Part of the estimates may be economically measured in indices similar to designed ones and used together with them (measures for removing negative consequences or using an additional benefit from the decision alternative under discussion may be planned). Estimates of individual processes development, which cannot be expressed in economic indices at the present state of knowledge, may appear (Fig. 1). However, these indices should be also considered in decision making; in our opinion, this is the main distinctive feature of the solution of the problem and its difficulty.

Proceeding from all accomplished designs and plans for interregional water transfer and for water supply of the national economy, and taking into account the earlier suggestions for solution of water problems, one may distinguish four quite different approaches, but each of these alternatives solves the same problem – the water supply of the national economy and the development of the country's water resources for a long-term period of over 40–50 years.

The first alternative (Fig. 2) envisages separate water supply of the European and Asian parts of the USSR. The water deficit of individual regions in the European area of the USSR should be eliminated mainly at the expense of the run-off of northern rivers of the Kara and Beloye (White) sea basins and lakes of the north west of the USSR which should be diverted

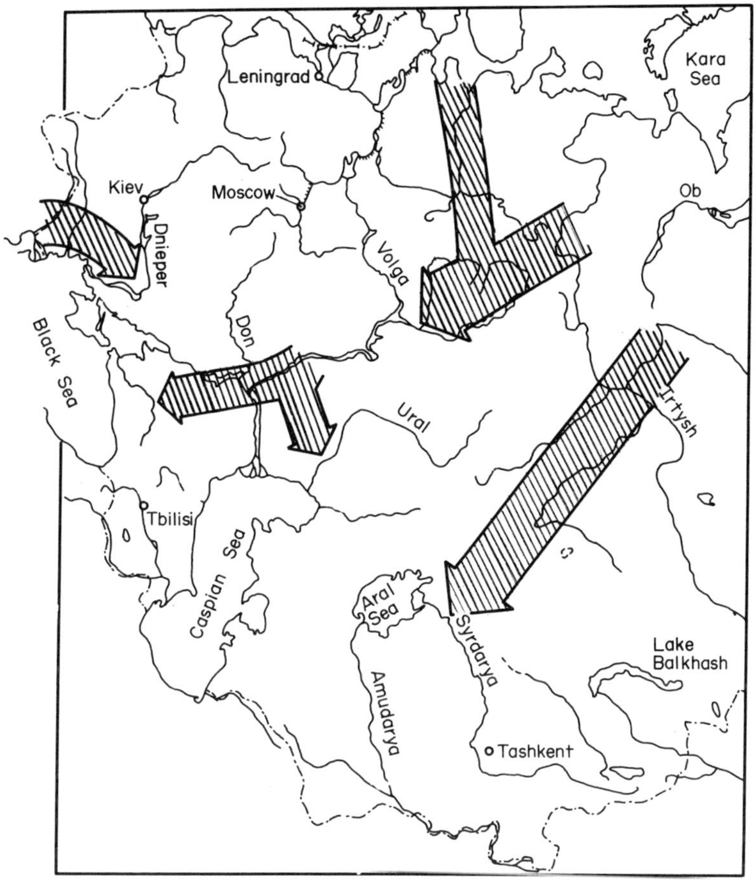

Fig. 2. Separate Euroasian alternative.

to the Volga. The diversion of the run-off of northwestern rivers and the Danube streamflow to the Dnieper basin will also contribute to the solution of the European area's water problems. In the Asian area of the USSR, the growing water requirements of Central Asia and Kazakhstan will be met mainly by the Ob River run-off.

The second alternative (Fig. 3) envisages the combined solution of water problems in the European and Asian areas of the USSR. The main idea consists in withdrawing water from the Lower Ob and transferring it through the Urals to the Pechora basin and farther to the Volga basin. The water of some northern rivers and lakes of the European area should also be diverted to the Volga. The water supply to Soviet Central Asia, Southern Kazakhstan, Middle and Lower Volga areas, Northern Caucasus, Kalmyk area and the Rostov region will be provided by the run-off of northern rivers, diverted to the Volga. The diversion of the run-off of the Danube and northwestern rivers to the Dnieper is also being contemplated.

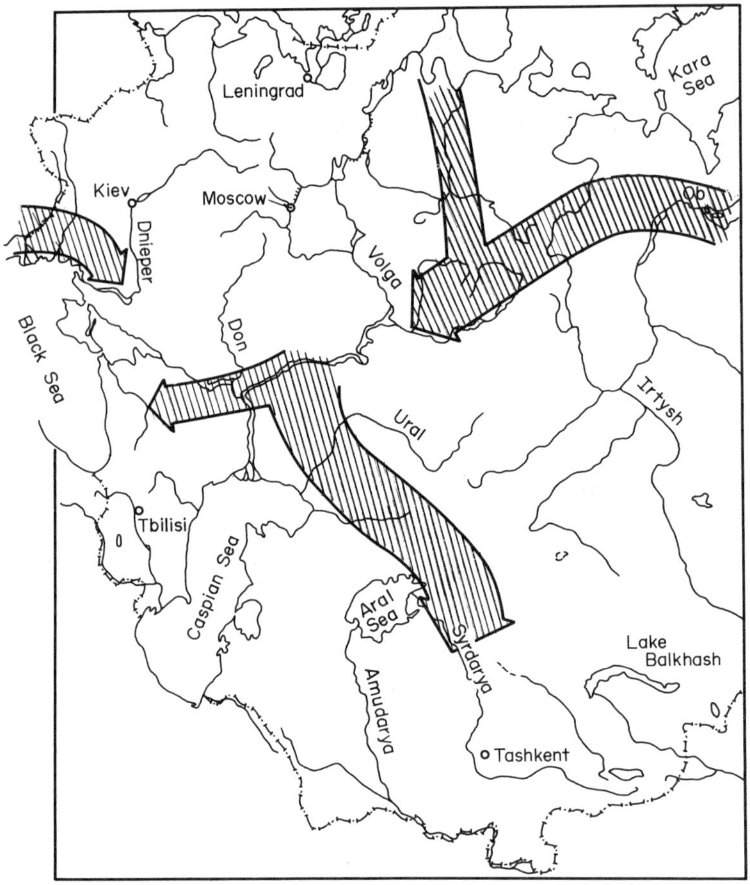

Fig. 3. Integrated Euroasian alternative.

The third alternative (Fig. 4) envisages the water supply of both the Asian and European areas of the USSR at the expense of the Volga run-off alone. In the Northern Caspian Sea a dam will be constructed, this will result in a manageable water–salt regime in this area. The Black Sea water will be transferred to the Caspian Sea. Local solutions involving the diversion of the run-off of other rivers may also be made.

The fourth alternative (Fig. 5) may be considered as a totality of all possible solutions to be made in an optimal combination within the Integrated Water-Resources System of the country. This last alternative represents the state of the nation's developed water management system and may be considered as the most general solution of water problems. The three above-mentioned alternatives may represent separate stages of the development of the Integrated Water-Resources System whose links may be formed completely in a longer period of time lasting for 30–40 years.

Now, it is possible to formulate a number of principles which will determine the general features of the Integrated Water-Resources System, its formation, stages of its development and functioning. They are as follows:

(1) The development of the Integrated Water-Resources System and the extent and quality of water resources management will depend upon the scientific and technological progress of the nation and the general state of productive forces. The rates of power systems development and river water power use are very important in this respect.

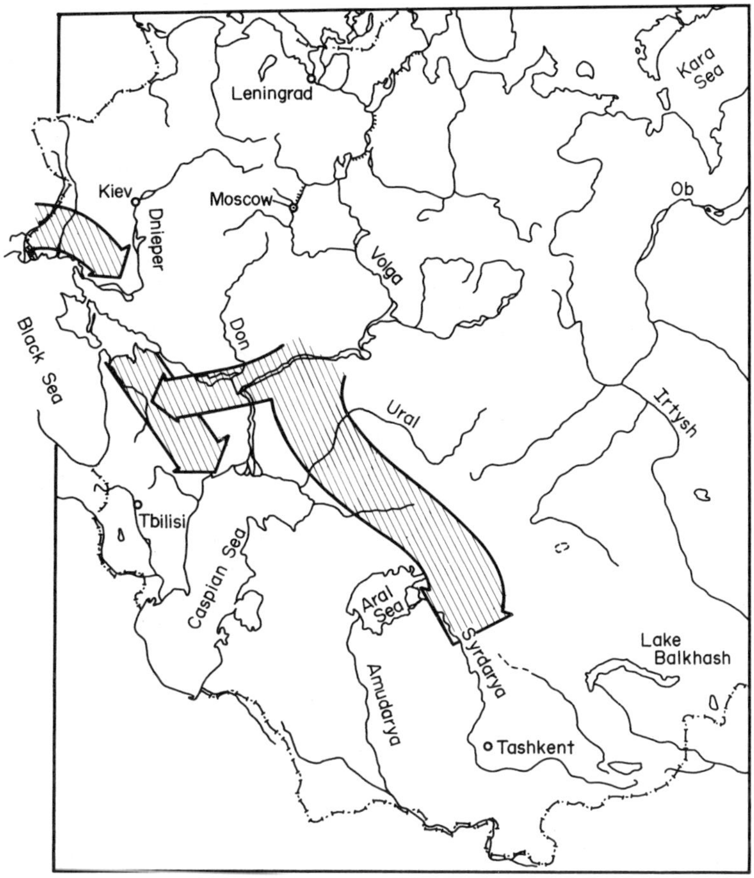

Fig. 4. Black Sea—Caspian Sea and Volga alternative.

(2) The Integrated Water-Resources System should be considered not only as a technological means of water resources management, but also as an important component of the environment, providing a basis for the formation and development of hydrologic, hydrochemical, ecological and other links.

(3) Under conditions of annual and long-term variations in water resources and water consumption, the water resources links being formed within the Integrated Water-Resources System should consider most completely the asynchronism of water resources and water consumption from various sources in different regions of this country.

(4) Proceeding from particular conditions of water resources distribution over the country's area, run-off regulation is advisable to be carried out not only by the routine method of the construction of inland water-storage reservoirs, but also by creation of regulating storages in river-mouth areas, using sea bays, sea areas and even individual small seas, for example, the Sea of Azov and the Beloye (White) Sea.

(5) It is advisable to assess any water management measure or development stage planned for the near future with regards to their place in the prospective Integrated Water-Resources System taking into account the general principles of the formation and development of the System.

(6) Selection of the next stage of development of the Integrated Water-Resources System should allow the possibility of further multi-variant development of the System. This will permit improvement in forecasting the current stage impact on the environment and economy and will allow a complex of studies for further development of the System to be carried out.

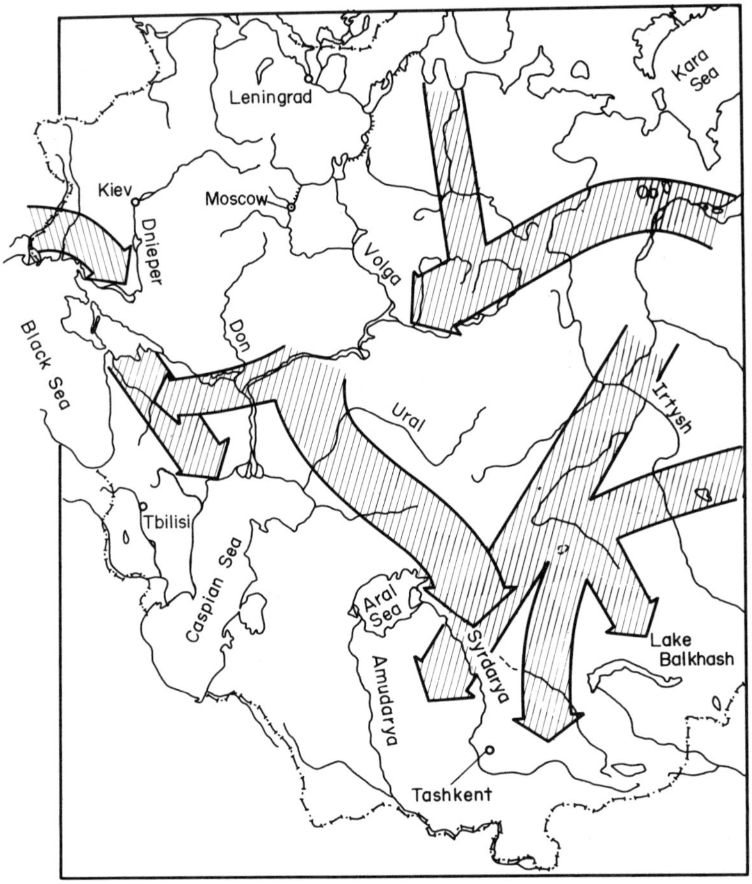

Fig. 5. The Integrated Water-Resources System.

The scope of the problem to be solved may be seen from the following figures: the total area within which run-off will be transferred amounts to 12,000,000 km^2 that is more than half the USSR area and more than all Europe, this area includes 600,000 km^2 of lakes and 1,000,000 km^2 of adjacent seas. The run-off of the area is 2200 km^3/yr or more than half the total run-off of the USSR, exceeding the run-off of the conterminous United States of America or that of India.

Each of the above mentioned alternatives provides equal water amounts, but may produce a different effect on the use of natural resources, development of productive forces, and the dynamics of environmental processes. Some of the effects are seen even now or may be clarified in the course of designing and planning, some others (and they are the majority) need special studies and prediction. For example, it is evident that the first alternative measures (separate European and Asian water diversions) may appreciably reduce the water discharge of the lower reaches of northwestern rivers in the European area and the middle reaches of the Ob in the Asian area of the USSR. The second alternative measures (the integrated Euro-asian alternative) are more preferable in this respect.

The advantages of the second alternative consist in stage-like character and smaller terms of funds freezing in the course of construction. For example, after the construction of the Volga–Aral Sea canal providing water for the Syrdarya and Amudarya basins, it will be possible to start construction operations for diversion of water from the north to the Volga basin. The water of the Pechora, Northern Dvina, northwestern rivers, and the Lower Ob may be used in various succession and amount, or the third alternative, comprising the diversion of water from the Black Sea to the Caspian Sea, may be developed. It may be evident that the transfer of large water amounts to the Volga will improve the quality of the water both in its reservoirs and the river run-off in the basin.

On the other hand, it is also evident that the transfer of the Lower Ob water over the Urals Range and permafrost areas is difficult. Diversion of water from the Black Sea to the Caspian Sea and regulation of the salt regime of the Northern Caspian area seem to be technologically complicated due to the necessity to uphold the salt regime of the Sea of Azov and due to the routing of the canal through areas with a high level of economic development.

The fourth alternative, involving an optimal combination of individual technological decisions, reasonable terms and scope of operations, may exclude some drawbacks of the other alternatives.

An appreciable advantage of this alternative is the possibility to create the Integrated Water-Resources System providing water resources management on a completely new basis, i.e. creation of regulated water bodies in river mouth areas and a possibility of manoeuvering water resources over extensive areas with allowance for the asynchronism of water resources and consumption.

Natural regulating storages in the river mouth areas of the Kara, Barents, White and Baltic sea basins (Ob and Yenisei estuaries, some bays of the White and Baltic sea) and the diversion of water from the storages to the south will preserve as much as possible the natural hydrological regime of river systems and reduce the need for reservoir construction on plains. Climatic conditions of the north would guarantee the quality of water resources. Acceptable conditions for fisheries and sea navigation would be created by regulating the water exchange between the water bodies being constructed and the seas. A similar approach to water resources regulation in the south in the basins of the Caspian Sea, Sea of Azov, and Black Sea may result in creation of storages with a regulated salt regime providing for intensive fisheries under conditions of rich biogenic run-off and heat abundance. This may be accomplished by the regulation of the Sea of Azov, Northern Caspian Sea and Black Sea's firths.

Within the Integrated Water-Resources System, the most rational management of water resources may be achieved by the diversion of water from one basin to another having different water demands.

It is known that water availability and demands within a basin may be correlated, but more often they are not. Hence, reasonable reallocation of water resources between river basins in different years should be performed.

Studying and predicting the effect of the planned measures are advisable to carry out on the basis of modelling the water resources system as a whole, and individual environmental and economic processes associated with the development and functioning of the Integrated Water-Resources System. At present, the experience in development of such major scientific problems and, the more so, modelling of such water resources systems exist neither in the USSR nor abroad. Projecting the Integrated Water-Resources System cannot be identified with any single mathematical problem due to the diversity of criteria. In this case, a series of problems connected by a chain of non-formal links inevitably arises. The selection of a solution variant must finally result in search for a certain compromise. Based on systems analysis, one should assume that the most effective way of solving such problems may be the use of a system of interrelated and subordinated mathematical models (Fig. 6).

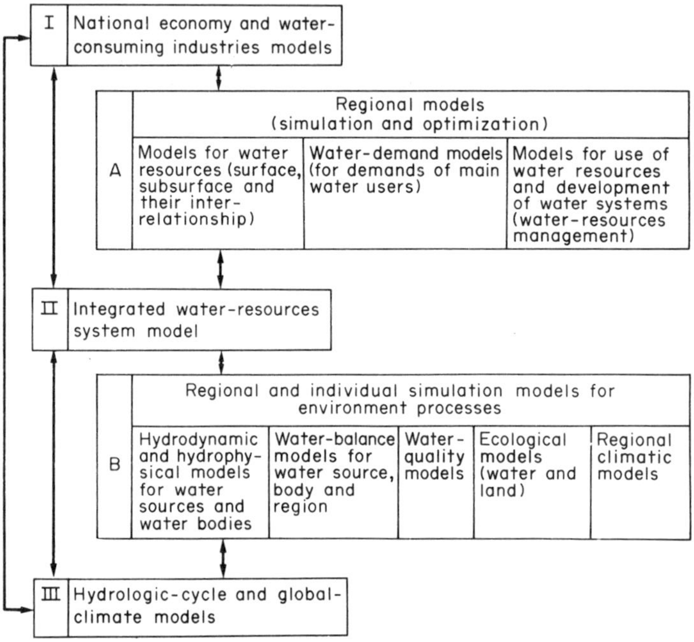

Fig. 6. A system of mathematical models for studying and estimating interregional water transfer.

A model, having input information on the location and distinctive features of the regime of water sources, location of water consuming industries, regime and requirements for water resources of all water users, and location and characteristics of water resources installations, may produce output data on the regime of water resources installations (management knots) and water bodies.

The information obtained is initial for development and solution of regional models of various environmental processes (Fig. 6, Block B) and global models (Fig. 6, Block III). On the other hand, the results may be used for correcting initial locations of water consuming units (Fig. 6, Block I) and management regimes (Fig. 6, Block A).

Proceeding from the experience gained for today and achieved level of programming, it is possible to realize the above problem, if the necessary information is available from special studies of environmental processes and a complex of design studies.

The starting point in these studies must be the prerequisites of development of water consuming industries and requirements for water resources of all branches of the national economy and water management.

Distinguishing water consuming industries from industrial branch plans through needed product balances and consideration of other demands on water will allow the formulation of the problem of locating water consuming industries and the construction of optimization models for the whole country. At present, an appreciable experience in construction of such models has been gained (Fig. 6, Block I).

Regional models (Fig. 6, Block A) are of essential importance in the system of models. Simulation models of the water resources of individual surface and subsurface water sources, models of the surface water–groundwater interrelationship are used for estimating the regularities of water resources formation and water source regimes with allowance for natural processes, and current and prospective economic activities. Water demand models reflect the regularities of formation of water demand in different branches of economy and take into account natural processes and technological development in water consumption.

Models of using water resources in separate regions and development of water systems may be optimizational. At present, an experience in numerical solution of such models exists for a number of river basins (Terek, Don, Kuban, and Syrdarya river basins). Linear and dynamic (both stochastic and deterministic) programming models are being used. Appreciable achievements in this field have been made in the USA.

The solution of problems using the Block A models allows the information on water resources, water demand and water resources management to be obtained. This information in the most aggregated form (water resources and demand of preset probability, run-off regulation parameters, economic indices of industries, etc.) is necessary for Block I models. The information from Block A models in a non-aggregate form is used in models of the Integrated Water-Resources System (Block II). In this case, indices of the dynamics of the natural water resources of river systems for a long-term period with their annual variation may be used, and water demand may be characterized by qualitatively identical indices.

In addition to the information role, Block A models should perform the correction role: they allow correcting water resources, water demand and water management parameters by iterative calculations.

Regional and partial simulation models of Block B are of particular importance in this system. They should cover a wide range of natural processes: hydrodynamic (water flow in river channels, canals, water reservoirs and hydraulic structures), hydrophysical (flow of sediments and admixtures, river bank and bed scour, the thermal regime of rivers and water bodies), hydrogeological (groundwater flow and discharge, storage of groundwater in aquifers, interrelationship of surface and groundwater and artificial structures) water balance (water balance of water sources and regions), water–salt balance (water–salt balances of soils, areas, and water bodies), water pollution and quality (quality of surface and subsurface water sources and bodies), ecological (ecology of water and land systems and their interrelationship) and climatic processes (water circulation, heat regime and evaporation).

Complicated links exist between Block B models, they are caused by links between corresponding processes. Data on location of water management installations and the regime of their operation, i.e. the regime of water courses and reservoirs and water consumption are initial for all models. It is necessary by means of Block B models to estimate a fairly large number of alternatives of functioning the Integrated Water-Resources System and make iterative corrections in the initial conditions of the models of Blocks II and I. Therefore, unification of Blocks B and A models seems to be an important matter. Typical models having a certain level of aggregation of input and output data can be constructed (the topological scheme of models will have, naturally, regional features).

The above system of models is the basis of studies in this field. Iterative calculations will allow combined estimates of various alternatives of interregional water transfers and stages of development of the water resources system and correct design solutions to be obtained. As the development of water resources systems is associated with long term construction and with even longer periods of environmental changes one must predict development of these processes for many decades in advance. In this connection, while solving the given problem, the levels of water demand at some stages, approximately for the years 1990, 2000 and 2030, closely related with national economy forecasts, are being considered.

At present, in the USSR, integrated studies on the interregional water transfer problem are being carried out. Over a hundred research organizations are taking part in these studies and the USSR Academy of Sciences guides them. The problem is treated as ecological and economical and as an inseparable part of the general problem of rational use of natural resources in the USSR. Therefore, its solution should serve not only economic objectives, but should also be a considerable contribution to the scientific substantiation and development of measures for increasing the efficiency of environmental quality management in the USSR.

The programme envisages revealing all possible main alternatives of the solution of the problem and predicting the reaction of natural processes caused by the possible interference with the water regime of regions and separate water bodies. Such an estimate must be made for all alternatives at various scales of interference and for distant future. The reversibility of biospheric changes should be estimated, the qualitative and quantitative aspects of changes from the natural and historical point of view should be determined, and the benefit and detriment of these changes for man should be evaluated.

Proceeding from the above, the following studies are planned:

(a) identify and estimate, scientifically and substantiatedly, all the ways of meeting prospective water demands (alternatives of water resources development);

(b) assess the effect of water resources development on the environment and socio-economic processes;

(c) work out the scientific principles of water provision and put forward primary tasks for planning and designing;

(d) make an all-round scientific assessment of the planned primary measures and substantiate measures for preventing and eliminating negative consequences.

The fundamental results of natural and social sciences will be used, and new developments in some branches and trends of these sciences would be achieved in solving these problems.

Particular attention is being paid to the studies of the processes of circulation of atmospheric, soil, surface and subsurface waters, and also waters of snow packs and glaciers, formation of land water quality under conditions of changing water exchange and economic activities. Based on these studies and achievements of biological sciences, it is necessary to develop methods for both quantitative and qualitative estimation of the impact of the water factor and man's economic activities on the ecological systems of land and water bodies, and therefore, all the environment. These studies are expedient to be linked with studies on the theory of climate for the purpose of a more complete and sufficiently substantiated estimate of man's impact on the environment.

Studies on scientific prediction of the development of the national economy for the distant future and prediction of scientific and technological progress in water consumption and water diversion are planned for solving the socio-economic aspects of the problem. The principles and methods of systems analysis are planned to be used widely both in posing the problem as a whole and in solving it. Methods of modelling large systems of physio-geographical and ecological processes, governed by the water factor, and modelling of the management of complex water resources systems will be used in particular.

The methodical basis of the programme is composed of present-day methods of analysis of complex systems with probabilistic and uncertain information. The objective of studies – the development of methods for water resources management allowing rational use of natural resources and environmental conservation – may be achieved only on the basis of all-round qualitative and quantitative analysis of decisions made in keeping with systems analysis principles.

The systems approach to the problem of rational use of water resources calls for development and wide application of mathematical modelling and programming. Simulation models permitting a wide range of computer runs with models of water resources systems, which cannot be treated analytically, occupy a prominent place in the studies. A water resources system as the main link in solving the problem of the provision of the national economy with water, is an interacting, hydraulically related totality of economic and ecological elements. Considering the peculiar role of water in the man–environment system, the programme of studies proceeds from the following premises:

(a) the necessity of treating water problems in their unity at a national level. Such a unity and interrelationship of regional problems with the general problem are stipulated not only by economic, but, above all, natural processes;

(b) the interrelationship of the effect of patterns of water resources use on natural processes not only in the area of planned water systems, but also in adjacent regions (heat and mass transfer of continental and oceanic waters, water circulation in the atmosphere, interrelations of ecological systems, etc.);

(c) the necessity of considering the developing water systems for a long-term period of no less than 40–50 years, several stages associated with long-range plans for development of productive forces – 1990, 2000 and 2030 being distinguished;

(d) the necessity of the development, analysis and comparison of different alternatives of productive forces development as a basis for the formation of water demand alternatives, water availability and environmental protection problems. Working out of such alternatives should consider direct and reverse links between productive forces and the water factor;

(e) the expediency of development of scientific hypotheses of water provision, founded on the scientific and technological progress and technology of water consumption, achievements in the management of water resources formation, the total potential of productive forces, and the possibility of construction of large-scale engineering projects for water resources management.

In the USSR, there is a number of objective prerequisites for successful solution of the problem of interregional water transfers for the benefit of all the people: state-wide planning of all economic and social activities, state property of land and water resources, high economic potential and vast experience in realization of large-scale water resources projects in irrigation, hydropower construction and water transport.

REFERENCES

Voropaev, G.V. Tasks and organisation of studies in connection with the problem of water transfer. *Vodn. Resur.* **3**, (in Russian), (1976).

Voropaev, G.V. The USSR Integrated Water-Resources System. *Vodn. Resur.* **6**, (in Russian), (1976).

Water Supply & Management, Vol. 2, pp.103-115. 0364-7714/78/0601-0103$02.00/0

Large-scale Transfers within Master Water Planning in Mexico*

HÉCTOR GARDUÑO, EDUARDO MESTRE and FRANCISCO TAPIA**

INTRODUCTION

Mexico, a country with a surface of nearly 200 million ha, has distinguished itself by an accelerated demographic growth, and an unequal spatial distribution. In 1950, the country had 26 million inhabitants, of which 60% were rural. In 1975 the population of Mexico reached 60 million, of which 40% lived in rural areas. It is estimated that by the year 2000 the total population, demanding great amounts of food from agriculture, will be from 126 to 139 million. Migration of peasants, caused by the hope of finding better living conditions in the city, will mean that by the turn of the century only 20% of the population will be rural and this will increase the size of already gigantic Mexico City and a few other large urban centers, which will require an increase in their water supply systems.

Agriculture is responsible for 95% of the total consumption of water. Although urban–industrial activities withdraw and consume small volumes of water, compared with those used for irrigation, the irregular distribution of water in the country and the increasing marginal cost of augmenting water supply systems, calls for huge investments to have adequate water supplies for our cities.

The average yearly rainfall in Mexico is 780 mm, equivalent to 1,530,000 million m^3; about one fourth of this corresponds to surface run-off, that is 410,000 million m^3.

The distribution of water resources has no direct relation to the location of the population. The southeastern area of the country, which contains 15% of the total area of Mexico and 12% of the population, has 42% of the run-off. On the other hand, the central and northern plateaus have 36% of the nation's territory, 60% of the population and only 4% of run-off.

The future need of water for irrigation and of huge water supply systems within a territory with non-uniform and non-coincident distribution of agricultural soil, water and population generate problems which will be treated in this paper. These problems, along with others such as water pollution, groundwater mining, etc., called for an integrated approach to water resources planning. A group was formed in 1972 to develop a National Water Plan, whose objective was to formulate and institute a systematic process for planning the water resources development for the rational selection of programs, projects and policies on this subject, which contribute to the attainment of the objectives of national socio-economic

---

* This paper was edited with information from the Mexican National Water Plan 1975 and ongoing projects of the NWPC. The part dealing with water transfers to supply Mexico City, was taken from studies of the Comisión de Aguas del Valle de México (Water Commission of the Valley of Mexico).

** Comisión del Plan Nacional Hidráulico de la Secretaria de Agricultura y Recursos Hidráulicos. (National Water Plan Commission (NWPC), Agriculture and Water Resources ministry).

development. In order to utilize foreign expertise and to make it possible for other countries with similar problems to take advantage of this effort, an arrangement was made with the United Nations Development Program and the World Bank.

A methodology was designed and a first iteration of the National Water Plan (NWP) was produced in 1975. In May 1976 the NWP Commission was created in order to update the Plan every 2 years, to constantly maintain in force its execution, to evaluate its results and to promote and coordinate research and training programs for water resources development. The NWP 1977 is presently being integrated.

METHODOLOGY, REGIONALIZATION AND SOME RESULTS OF THE NWP 1975

Methodology

The socio-economic aspects, as shown in Fig. 1, are studied at both national and regional levels, allowing in this way an identification of the objectives, policies and development goals which, along with technical factors, determine the demand for water. The supply of water resources is compared with the demand. In this way, the balances are computed, and problems derived from water shortage and lack of control or inadequate quality are identified. The alternative solutions to these problems, proposals made by different agencies related to water, and a catalogue of existing projects, are the basis to formulate and integrate, with the aid of systems analysis techniques, the programs for water resources development in each of the regions in the country.

The regional programs are integrated at national level in order to analyze the compatibility of the supply achieved through the programs with the demands derived from development, and of the financial and human requirements, with their availability. Through this analysis, some adjustments are made in goals and policies originally formulated; in this way, a cycle of the planning methodology is completed. The results of the implementation and operation of these programs set up a feedback with the socio-economic scenario, and lead to the identification of new goals.

In each cycle of the process, objectives, policies and goals for water resources development are set up. New projects are identified and recommendations are made for making studies and obtaining basic data. Finally, programs to build infrastructure works, provisions for its operation, institutional modifications, personnel training and research are formulated.

Regionalization

The hydrological watershed is the most adequate unit for planning water resources development since it groups, in natural form, the diverse groups of users and permits an integrated consideration of the effects of the management of water. For this, the country was divided into thirteen regions, formed by the main hydrological watersheds and groups of watersheds. These regions were grouped into four zones, each one including areas with similar characteristics or related in some way through the use of water. The thirteen regions were then divided into 102 socio-economically and politically homogeneous subregions. In Fig. 2, the division of the country into zones, regions, and subregions is shown.

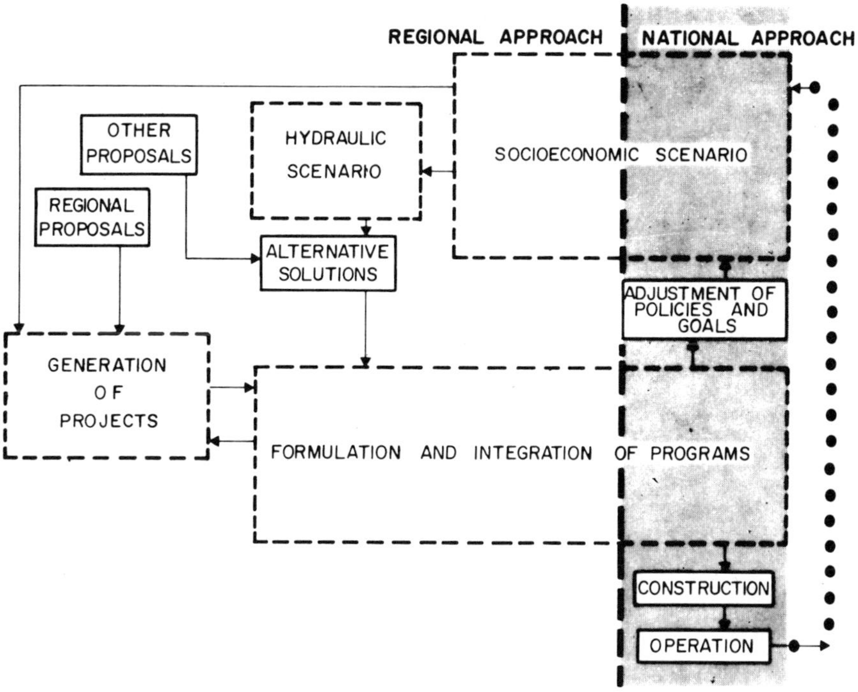

Fig. 1. Methodology of the studies of the National Water Plan.

The National Water Plan 1975

The dynamic process described above resulted in the NWP 1975 report, which was divided into three parts, as shown in Fig. 3. The first part is composed of the socio-economic and physical frameworks as well as the regional strategy for water resources development which is described below. The second part included the diagnoses, objectives, goals, policies and programs for water resources development, for each of the different activities which use water.

The third part is the integration of the program and policies; an analysis is made of its financial, technical and human resources feasibility, and the actions necessary to implement the Plan are suggested.

Regarding irrigation and drainage, construction and improvement of the hydro-agricultural infrastructure will contribute to the solution of some of the problems faced by the farming and livestock sector, since this permits an increase in the productivity of land, generation of employment, and an increase in the aggregate farming and livestock value.

The internal demand for farming and livestock products has registered an accelerated increase due to the increase in population, *per capita* income and inputs for industry. Until recently, it was possible, on the average, to satisfy basic needs and to export articles which permitted financing for the development of the nation's industry. However, there are still serious nutritional problems due to the low income level of a large part of the population and the lack of knowledge regarding proper diet.

At present there is an area of almost 5 million ha with hydro-agricultural infrastructure; these hectares are distributed in irrigation and drainage districts, irrigation units for rural

Fig. 2. National Water Plan's zones, regions and subregions.

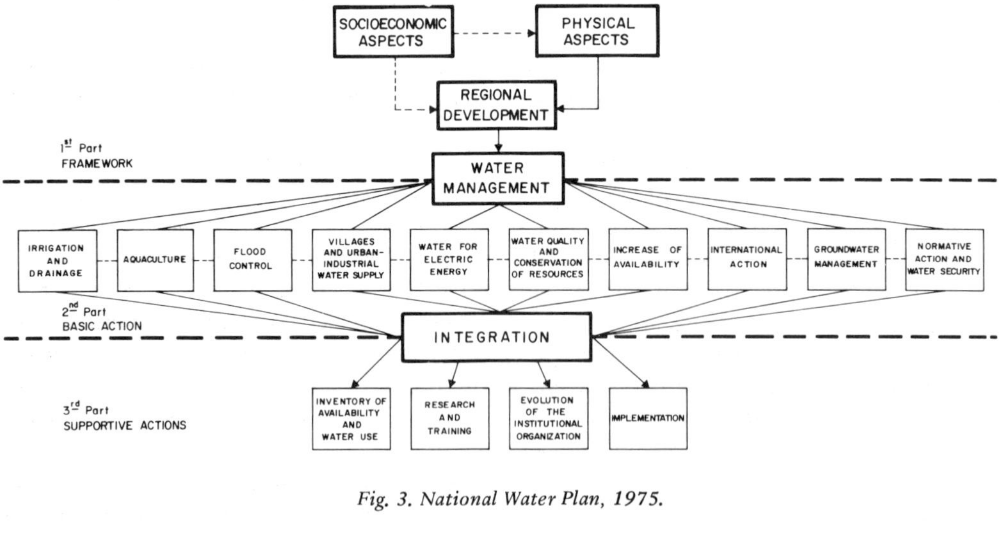

Fig. 3. National Water Plan, 1975.

development, and in private property, which represent 30% of the national harvested area, contribute 50% of the value of the total agricultural production, and contribute a large variety of basic products.

The goals proposed for the year 2000 involve the duplication, in 25 years, of the area with hydro-agricultural infrastructure put in operation during the last 50 years, reaching 10 million ha by the end of the century. The following paragraphs describe the suggested strategy for the regional water resources development of Mexico. In order to meet our agricultural production demands, large-scale water transfers are needed in the Northern and Central Pacific Zone (NCPZ). Mexico City, in the Central Zone (CZ), will increase the use of water from surrounding watersheds.

Regional strategy for water resources development

The soil and water inventory, the computation of water demands and potential pollution in different uses and for different alternative scenarios, and identified projects, allowed the accomplishment of integrated water and soil balances. This study required the use of detailed models of hydrologic simulation at watershed level, and estimates concerning the possible effects of pollution.

The results of these balances, expressed as water and soil potential, are shown in Fig. 4.

In the *Northern and Central Pacific Zone*, water resources and soil with high agricultural potential, part of which is still uncultivated, do not coincide spatially. There are 1.5 million ha unexploited, the majority of which are located in the central and northern part of the coastline of the Northwestern region, while abundant water is found mainly in the Pacific Central region.

Irrigation infrastructure has played an important role in economic development. Those areas under irrigation can still grow considerably larger with new hydraulic development; this can be accomplished through transfer between watersheds and local water resources developments, and in a very important way, through an increase in efficiency, taking advantage of the aptitude of the production apparatus of the area to increase its production and respond to the incentives without much inertia.

Under these conditions, the increase in harvested area could be significant. In spite of this, irrigated agriculture is not considered to be the only activity capable of satisfying the great demand for jobs which the increase in population will provoke; the great need to motivate the industrial, agro-industrial, and service sectors stems from this. For this expansion, it is advisable to reinforce some urban centers and industrial corridors located between the northern border and the country's central economic system.

It will be necessary to provide these centers with an infrastructure of water supply and sewerage, services which will also be required because of the encouragement of tourist activities, mainly in Lower California.

Aquaculture in fresh and brackish water is another activity which has great importance both in food and inputs production for industry as well as in employment generation. This activity has received a great deal of incentive since 1970; there is still a great potential which is extensively and intensively exploitable along the coastline, in present and future reservoirs, and through rural fisheries.

In the *Northern Zone*, geographic conditions dictate an extreme, arid climate in a large part of the region. It has the lowest precipitation in Mexico. The increase in irrigated areas has

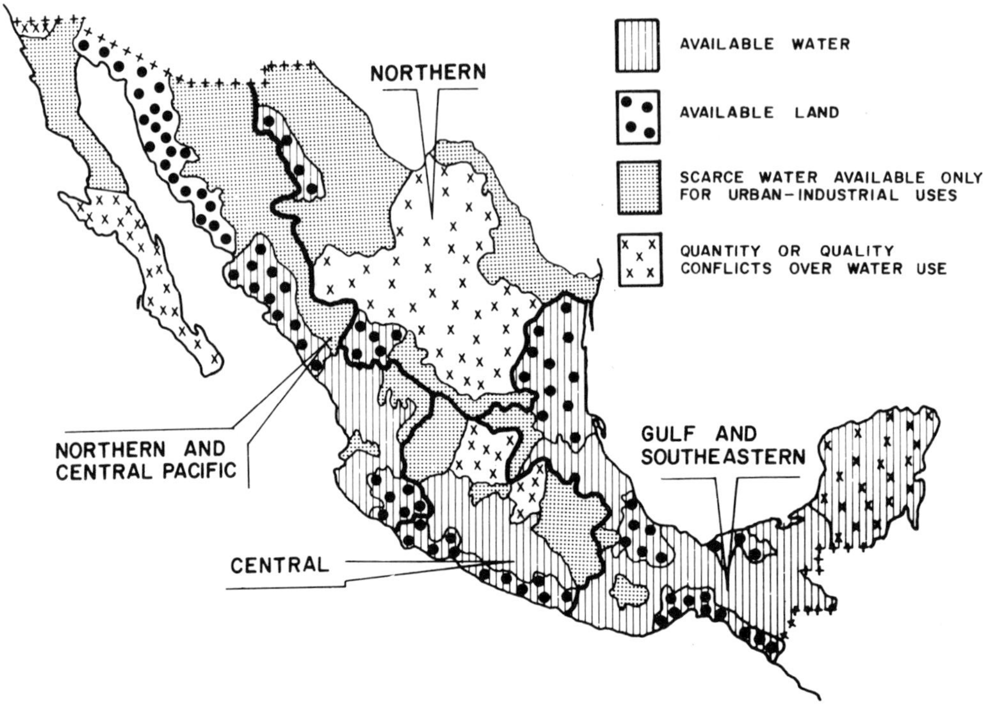

Fig. 4. Water and soil framework for the regional water resources development.

become stagnated due to the scarcity of water shown in Fig. 4. In the present decade, the increase in the work force is absorbed by a dynamic secondary sector; this in turn causes a greater urban concentration. Together with the problem of little availability of water, this resource is managed with low efficiency, especially in agriculture. In general, investments oriented towards efficient water management are highly profitable because of the opportunity cost caused by water's own scarcity. This region does not compete with the other three for government investment grants to open new areas through big hydro-agricultural projects, because of the small reserves of water still available. For this reason, the investments are mainly oriented towards projects tending to increase productivity.

In an increasing measure, the industrial sector is the one which absorbs the increase in the work force. The expansion of this sector is limited by the water scarcity; this, along with characteristics of the regional industrial profile, support the idea of selective growth with increase in productivity and a more efficient use of water which considers the reuse and recirculation of this resource. Water supply for the large urban industrial centers will come to a greater and greater degree from groundwater sources, except in a few cities, supplied from the River Grande, whose future growth will affect the hydro-agricultural areas which at present use water from the same river.

The most outstanding and attractive physical characteristic of the *Central Zone* is its climate; this has, to a degree, originated the immigration to this area, especially to the Lerma and Valley of Mexico regions, and to the high part of the Balsas River watershed.

This region has some very peculiar characteristics such as an agricultural tradition dating from the prehispanic era, disproportional distribution of the population and of economic activity in the metropolitan areas of Mexico City and Guadalajara, a net of cities of secondary importance with more than 50,000 inhabitants, and a serious income inequality between urban and rural areas.

Water resources and the demand centers do not coincide geographically; more than 80% of the run-off is formed in the Balsas region, where, because of its accidented topography, it is difficult to accomplish significant new hydro-agricultural development. On the other hand, the intensive use of water in the Lerma and Valley of Mexico regions causes conflicts due to the fact that the availability of water in these areas is not sufficient; this problem is made worse by the intense and growing pollution of the river courses, and in some cases, by the waste and loss of water in the systems.

Based upon these characteristics, the following global orientations for hydraulic development have been identified: to increase efficiency in the use of water, above all in agriculture, the main consuming activity, and to intensify the schemes for reuse of water in the urban—industrial conglomerates; to open new areas of agricultural production with irrigation, diversifying their crop patterns to participate in improved conditions in the consumer markets in the region; to decentralize industrial activities with basis on a selective location, taking into consideration the withdrawal, the consumption and the quality of the waste effluents, as well as the type of production; and to improve the level of the municipal water supply. Considering the growth and concentration of the population, from the years 1980 on, it will be necessary to make water transfers from nearby watersheds to the Valley of Mexico, maintain an adequate quality of water resources, as well as of aquatic organisms, and to implement a more efficient and coordinated water resources management.

Agricultural activity occupies the major part of the economically active population in the *Gulf and Southeastern Zone*; 96% of production is obtained in rain-fed areas.

The management of the scant hydro-agricultural infrastructure is very inefficient, mainly due to the characteristics of the soil, to the lack of technical assistance, and to the climate. The main checks on development are: from a technical point of view, the lack of experience in intensive agricultural exploitation in tropical areas, and from a social and cultural point of view, the scant knowledge of the motivational factors of the population, as well as the profound modification of the agrarian and social structure caused by the change from an individual farming and livestock production of subsistence level to an intense commercial production on a collective level.

The development of the vast natural resources of this zones, shown in Fig. 4, is a challenge which will have great economic importance for the country. Due to the difficulties mentioned here, the Gulf and Southeastern Zone would seem to be the least attractive, from a purely economical point of view, to channel federal investments for hydro-agricultural development; however, in order for the inhabitants to benefit from the national policy for shared development, and to satisfy future demands for agricultural products, the process of intensive production through adequate policies and incentives must be incorporated to the zone.

Since the agricultural technology and rural administration in the ecological conditions of the tropics are little known (Biswas, 1978), a strategy of development by stages must be implemented for hydro-agricultural projects in which an infrastructure oriented towards drainage, flood control and supplementary irrigation is considered. The first stage of each project is a pilot project whose main objective is to test modern technology and the corresponding administrative system which help the users to acquire the entrepreneur's capacity necessary in the new types of exploitation. Parallel to the building of structures, it is indispensable to develop and adapt technologies which are adequate for the humid Mexican tropics, to carry out specific farming and livestock research, to train technical personnel, and spread the results of the research and experiments to the users. The creation of an Institute of Tropical Development is foreseen, for the purpose of applying these investigations and experiments.

Some conditions which favor an important increase in industrial activity and in the infrastructure of this zone, as well as the opening of new work centers are: a commercial policy to dynamize the exports even more and diversify external markets, the recent trend to decentralize economic activity, the possibilities of farming and livestock development which will propitiate the formation of agro-industries, the enormous hydroelectrical potential and the discovery of new oil fields. In this sense, a dynamic development in industrial centers in the Northern Gulf and Southern Pacific Isthmus regions can be foreseen. Merida is another urban area of importance, though it is not as dynamic as those regions mentioned above; its location on the Yucatan Peninsula and its cultural tradition make Merida a city which should be integrated, along with future developments in the Mexican Caribbean, to an extensive area of touristic interest.

WATER TRANSFERS FOR AGRICULTURAL DEVELOPMENT IN THE PACIFIC REGIONS

A series of rivers crosses the coastal strip of the Mexican Northwestern region from the Santiago River to the Hermosillo coast; these rivers have a mean annual run-off of 26,500 million m^3. As can be seen in Fig. 5, 900,000 ha of this approximately 6 million ha area are now under irrigation utilizing annually 10,700 million m^3 of surface water from these rivers controlled by nine large storage dams and 1530 million m^3 of groundwater.

Along this strip there are still 1.5 million ha of idle lands adequate for irrigation agriculture; 80% of these are located in the northern part, while the rivers which are still uncontrolled are in the south.

The present development of surface water is being carried out through the conduction and distribution for irrigation of areas located near the rivers already controlled. At present, storage dams with a total capacity of 6300 million m^3 are being built; these dams will allow the incorporation of approximately 150,000 ha to irrigation through an interconnected system from the San Lorenzo River to El Fuerte River.

In order to develop 800,000 ha more, systems which would satisfy the demands of areas near each river and would allow exportation of excess volumes of water to the Northern region are necessary. Studies on water transfer for this region were begun 10 years ago. In the feasibility analysis, deterministic models of digital simulation have been used; these models reproduce the monthly behavior of water supply and demand centers, taking into account historical run-offs in the rivers during a 20-year period.

In the short range, the opening of some 50,000 ha has been proposed; this would be carried out by rational and temporary groundwater mining in the aquifers located between the Piaxtla and Yaqui Rivers for about 10 years. At the end of this period the groundwater will be substituted by surface water, and the surface being irrigated will be increased by 180,000 ha since the works for enlargement of the interconnected system between the Piaxtla and the Yaqui mentioned before will have been completed, as shown in Fig. 6. These structures include five storage dams with a total capacity of 6000 million m^3, a 240 km conduit and a distribution and drainage infrastructure for 230,000 ha.

The economic evaluations which have been carried out are favorable for the entire program over both short and medium ranges; that is, for the actions considering the rational mining of aquifers as a first stage followed by the enlargement of the transfer system from the Piaxtla River to the Yaqui.

A model of linear programming which works coupled with the hydrological simulation model has been used to define the scheduling of medium range works and the allocation of water in the Piaxtla–Yaqui system. The objective function considers net regional benefits and the linear model has as restrictions the available land and water in each center of demand and supply, respectively, as well as the production volume per crop which has been imposed at the regional level based upon the demand projections at the national level, and the participation which the northwestern region is estimated to have in the satisfaction of these demands. The results of the linear model are tested in the hydrological simulation model until a convergence between the results of both models is reached. The scheduling of the works is determined by ranking the values of the objective function obtained when different alternative components of the system are included.

This water allocation mechanism not only allows finding the best water distribution, but also makes it possible to find out the cost to the country of both alternative policies of water distribution as well as directed policies to stimulate the production of a certain crop in any of the production areas. This cost is estimated by calculating, for different alternative policies the decrease in the value of the objective function and the production volume which is no longer produced.

Over the long range, the integration of the great northwestern system from the Santiago River to the Hermosillo coast (shown in Fig. 6) is being considered; this implies another six dams, 1500 km of conduits which includes 12 km of tunnels and pumps and requires around 600 GWh/yr to raise the water 500 m. This system permits the opening of another 450,000 ha through the water exportation and of 130,000 ha in the vicinity of the controlled rivers, for which another 7500 million m^3 will be utilized annually. This volume of water will also make it possible to suspend the mining of the aquifers in Southern Sonora.

The long range program, which is planned to be completed in the year 2000, offers the opportunity to attain an important increase in the national grain supply, in such a way that our country be self-sufficient in those products. For this reason, traditional evaluation indicators are not given too much weight, and less so, considering the changing nature of the price system.

Since the water resources systems have multiple uses, the short, medium and long range construction programs lead to the opening of a total of a million new hectares to irrigation, to the generation of 3650 GWh/yr, to the provision of flood control in the coastal flatlands and to the regulation of the fresh water flow to coastal lagoons to propitiate aquaculture activities.

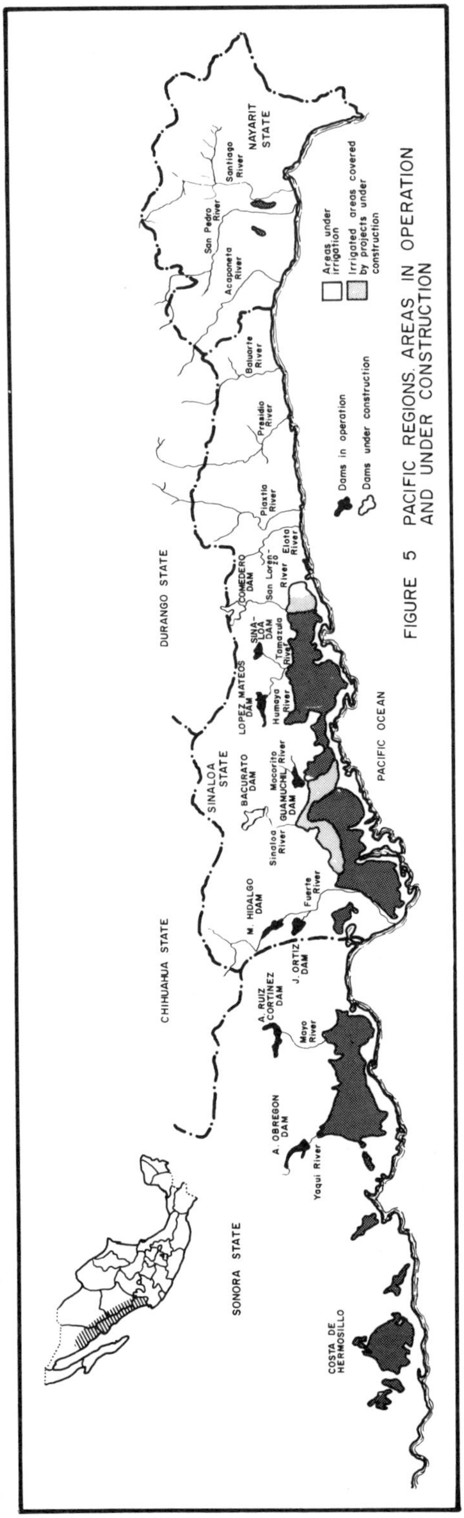

FIGURE 5 PACIFIC REGIONS. AREAS IN OPERATION AND UNDER CONSTRUCTION

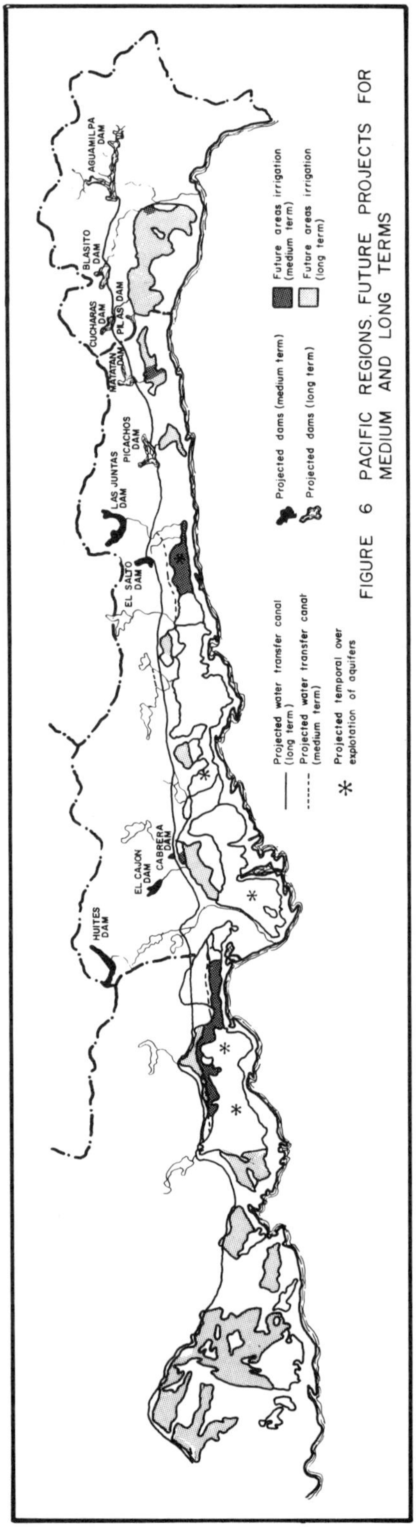

FIGURE 6 PACIFIC REGIONS. FUTURE PROJECTS FOR MEDIUM AND LONG TERMS

TRANSFERS FOR THE WATER SUPPLY OF MEXICO CITY

The metropolitan area of Mexico City, located in the Valley of Mexico in the central plateau of the Mexican Republic, 2240 m above sea level, and with a population of more than 10 million, now faces one of the most serious problems in the world regarding water supply.

The water supply problem has worsened due to the explosive way in which the population has increased (6% annual rate). At the middle of this century mining of aquifers underlying Mexico City was begun. This caused a notable increase in land subsidence and considerable damages, especially in the sewerage nets. From 1957 on, the importing of 13 m^3/sec

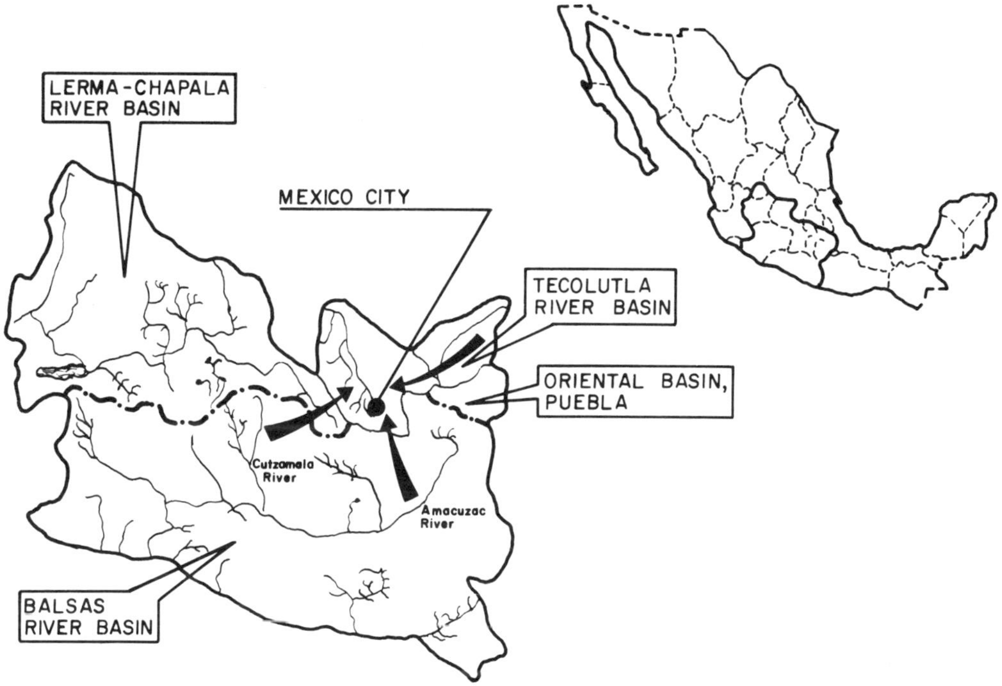

Fig. 7. Water transfer projects to the Valley of Mexico.

from the Lerma watershed, shown in Fig. 7, was begun. These aquifers are located about 70 km from the capital, and it was thought that this was the long range solution. However, now there are mining problems.

The water supply to Mexico City increased from 0.8 to 42 m^3/sec between 1900 and 1977. It is estimated that by 1980 the hydrological potential of the Mexico Valley watershed will have been used to its maximum.

The federal government is treating the problem of water supply to the Mexico City Metropolitan area through the Water Commission of the Valley of Mexico (WCVM), an agency created fundamentally with the idea of delivering the necessary volumes of water which are distributed in the corresponding networks by the municipal authorities. The programs of the WCVM will solve this problem up until 1979 by using the resources of the Mexico Valley watershed. It is estimated that by the year 2000 the water supply system will provide 109 m^3/sec – including the reuse of 5 m^3/sec – to about 30 million people in Mexico City. Until now, four alternatives have been identified: the Cutzamala River, the Tecolutla River, the Amacuzac River, and the closed watershed of Oriental, as can be seen in Fig. 7.

The WCVM has carried out technical–economic studies in order to analyze and evaluate the different alternatives. Until now, it has been demonstrated that the Cutzamala River is the alternative with minimum cost, even taking into consideration the consequences these transfers have in the agricultural and hydropower uses downstream from the places from which the water is physically to be transferred.

During the year 1976 the National Water Plan Commission carried out the Mexico State Water Plan project; a large part of the Cutzamala River watershed is located in this state. In this plan a study was made of the hydraulic feasibility of carrying out the water transfers of the Cutzamala River watershed, using a digital simulation model, which reproduces the behavior of the considered storages and considers different operational policies of these storages. The result of the analysis shows that it is feasible to transfer 19 m^3/sec in the first stage of this watershed with the existing hydraulic infrastructure, giving first priority to water supply for Mexico City. This implies a 46% reduction in the new agricultural project areas and a very important decrease in the hydropower potential in the plants which use the water to be transferred. It is estimated that, if carried out the transfer of 19 m^3/sec, the mean annual generation of all the affected plants would be reduced by 25%, which corresponds to 1500 GWh/yr.

Given the high opportunity cost of each m^3 of water for the metropolitan area of Mexico City, it is considered that the water allocation policy of this watershed should be for this purpose by means of water transfers and not for agricultural ends, since there are other possibilities for developing agriculture in the region, such as the technication of the rain-fed areas and irrigation with groundwater.

At present, the Water Commission of the Valley of Mexico is studying the evaluations which correspond to the second stage of the Cutzamala River watershed and to the Tecolutla, Amacuzac and Oriental alternatives in hopes of programming over a longer period the best solution to the problem of municipal water supply to the metropolitan area of Mexico City. Schemes for the future consider a strong component of water reuse and the improvement of efficiency in its use. However, all of the alternatives solutions require large investments and enormous quantities of energy for operation.

CONCLUSIONS

Large interregional water transfers must be tied to master water planning and to national global planning in a country. The main reasons to arrive to this kind of solution is the uneven and non-coincident distribution of water resources, soil, population and regional development.

In Mexico, the first national priority of food self-sufficiency will probably call for large water transfers in the northwest to take advantage of potentially productive arid land. The application of classical project evaluation criteria pose the problem of using unstable prices for agricultural products. Also, interregional water transfers imply the use of significant amounts of energy and, given the rising cost of this input, it will be harder to justify economically those projects in the future. A fresh look at existing methodologies and possibly formulation of new ones, is necessary so that projects that help to achieve national goals can be rationally scheduled.

On the other hand, the accelerated population growth and migration will make it necessary to devote large investments for the water supply of Mexico City. Greater efficiency in the use of present supply is urgent, as well as measures to reach a decentralized and better balanced regional development.

Since interregional water transfers usually involve several political entities within a country, watershed organizations are needed to achieve an effective water management that take into account political conflicts among the states covered by the hydrological regions. Also a central planning and management body is necessary to make sure that water development in fact helps to achieve national objectives and balanced regional development.

REFERENCES

Biswas, Asit K. Climate, agriculture and economic development in *Food, Climate and Man,* Biswas, M.R. and Asit K. (eds). John Wiley, New York, (1978).

Water Supply & Management, Vol. 2, pp.117-125. 0364-7714/78/0601-0117$02.00/0

Interregional Water Transfers: Case Study on India

K. S. S. MURTHY*

India is the seventh largest country in the world. Its territory extends over an area of 3,327,520 km^2 and the population is 618 million. Agriculture is the predominant occupation of the people and more than 70% of the people are engaged in agriculture. Agriculture contributes more than 50% to the gross national product. Successful agriculture in most parts of the country is not possible without irrigation or the artificial application of water to land. This is because of the peculiar climate that India and most parts of South Asia experience. The bulk of the rainfall precipitates in the monsoon months from June to September. Even during these months the rainfall in some places is more than is required by the crops but in other places is deficient even for the cultivation of ordinary dry crops. Also during the 'monsoon' months, sometimes the gap between two precipitations is so large that it affects the plant growth. Pressure on land is quite high and therefore, there is greater need for irrigation. Provision of irrigation affects the environment and life in a big way. To appreciate the problem of India, it may be helpful to recollect in brief, the geography and climate of the Asian continent.

Asia covers about one-third of the earth's surface but has nearly two-thirds of its population. The continent incorporates many different kinds of topography. The central areas are made up of plateaus and high mountain ranges. Great deserts extend in a wide strip from the ever-warm deserts of the Arabian peninsula northeastwards to the Gobi in Mongolia, where the winters are very cold. These desert areas are mostly inhabited by nomads who, like their counterparts in Africa, wander in search of water and pasture for their livestock. Northern Asia has extensive areas of tundra and coniferous forests. Further south, there is a steppe zone of grasslands and in the very south lies the hot and rainy equatorial zone with its steaming jungles. This part of Asia figures largely among the water-deficit areas of the world. Thus, rainfall is not sufficient for the type of vegetation which temperature conditions permit. This applies to the entire western part of southern Asia. India and Indonesia are the only places with a water surplus.

Apart from its mountainous areas, southern Asia has a hot climate all the year round. The year is divided into a rainy season and a dry season. The rains last from June till October, with daily cloudbursts in some places. Cherrapunji, in the north of India, receives more than 10,000 mm (10 m) of rain every year. The regular alternation of rainy and dry periods is connected with the monsoon winds, which are the main factor governing the climate of Asia. During winter, a dry, cold wind blows from Central Asia towards the southern and eastern parts of the continent whereas in summer, the wind blows in the opposite direction, carrying moist, rain-laden air over the land areas. The reverse applies in southwest Asia where the summers are long and hot and the winters mild. In some areas winter is the rainy part of the year, during which crops can be cultivated without irrigation.

* Chief Engineer, Central Water Commission, Government of India, Bikaner House, New Delhi 110002, India.

Most of the rivers of Asia rise in the mountain areas of Central Asia, whence they radiate – the Ob, Yenisey and Lena northwards to the Arctic Ocean; the Indus, Ganges, Brahmaputra and Mekong sourthwards; the Yangtse-Kiang and Yellow River eastwards. Millions and millions of people inhabit the warm, fertile valleys of these rivers. Needless to say, climate has a crucial effect on people's living conditions.

India is a continent in its own right, larger than western Europe. The country is a union of 22 federated States and nine centrally governed territories. It has the same disparity of languages, ethnic groups, religions and geographical subdivisions as Europe. The population of India is growing by about 14 million every year. Fifty five per cent of the total area of the country is cultivable, and agriculture accounts for half the national income. Climate, geology, topography, soil conditions and vegetation all vary a great deal; India has lofty mountain ranges, undulating hills, high plateaus, and rolling plains. Savannah and steppe are the natural vegetation of large areas of the peninsula.

South Asia derives the greater part of its precipitation from the southwest monsoon, which blows from the Indian Ocean. This monsoon is made up of two moisture currents. One, coming from the Arabian Sea, strikes the mountain ridge of the Western Ghats along the west coast of India, discharging large amounts of rain in the process. The other comes in over the Bay of Bengal, where it veers northwest and becomes a southeast air current sweeping over the lowlands of northern India. Both these air currents carry large quantities of moisture but they discharge only 20% of their moisture content on India. Spread out over the entire area of the country, this is equivalent to an average annual precipitation of 1100 mm, making 3700 km^3 in all. About a third of this precipitation evaporates and about 1700 km^3 returns to the sea through the rivers. It is estimated that the remaining 20%, or 790 km^3, seep into the ground and recharge deep-seated groundwater reservoirs. Of this amount, 270 km^3 are thought to be available. The densely populated valleys of the Ganges and Brahmaputra are particularly well endowed with groundwater. Central India is a primary plateau of rocks similar to those underlying large areas of Sweden and Finland. This rock is usually well-fissured, especially in the uppermost 30 m.

India's precipitation is unevenly distributed. In the eastern parts of the Himalayas and along the mountain range of the west coast it amounts to 4000 mm/yr. In the east of the country, local precipitation can be as much as 10 m/yr in parts of Assam. Central and southern India, on the other hand, lie in the rain shadow of the Ghats and receive less than 600 mm/yr, which is roughly the same precipitation rate as southeast Sweden. The driest areas are the northwestern States of Rajasthan, with its Thar desert, and Gujarat, north of Bombay, where precipitation is less than 100 mm/yr. Ninety per cent of India's rainfall falls during the short rainy season between June and September.

The rivers drain the entire area of the country, except for the desert area of Rajasthan, and for the most part run east and west. The following is a summary of India's annual resources of river water:

	km^3
Ganges–Brahmaputra	876
Indus (eastern tributaries)	40
Eastward flowing rivers in the Indian subcontinent (Mahanadi, Godavari, Krishna, Cauvery etc.)	414
Westward flowing rivers in the Indian subcontinent (Tapi, Narmada etc.)	308
Total	1638

It is stressed that these figures are not reliable because it is only recently that gauging has been undertaken on any considerable scale. There are two kinds of river: snow-fed, giving rise to perennial floods in the north and northwest, and monsoon-fed, associated with intermittent floods in central and southern India, where rivers regularly dry up during the dry season.

On account of its tropical climate, the irregularity of the monsoons and their limitation to a few months of the year, India is greatly dependent on irrigation for stable and successful agriculture. Without irrigation, farming in India would be a gamble with the forces of nature. The great plains of the Indus, Ganges and Brahmaputra have been farmed for at least 6000 years. Irrigation is also a very ancient science and there are many remains of prehistoric constructions. Some installations from historic but remote times are still in use today. Irrigation received a stimulus in the mid-19th century, but the partition of the subcontinent in 1947 gave most of the irrigated areas to West Pakistan. India acquired land areas where rainfall was highly capricious. The total irrigated acreage at that time amounted to some 20 million ha equalling about 20% of the cultivable area. A massive development programme was launched which gave top priority to the development of water resources. The irrigation acreage in 1947 has now been at least doubled. Altogether, 22 million ha are now provided with large or medium size irrigation works and an equally large area provided with small-scale works.

DEVELOPMENT TO DATE

As already mentioned, at the time of Independence in 1947, India had an irrigation facility of 20 million ha. Since then during the last 30 years more than 500 major and medium irrigation projects have been taken up and nearly three-fifths of them have been completed so far. Quite a few are in the process of completion and have already started giving benefits. Even so, the development of irrigation to date is roughly of the order of half the total irrigation potential estimated for the country.

Even the very process of estimating total irrigation potential of the country is still in a very preliminary stage. Most of the work is based on paper studies and assessments. Detailed investigations and preparation of basin plans are yet to be undertaken. However, one thing is clear from the studies carried out so far, that the water resources of India are distributed geographically in a very uneven way. There is plenty of water in the northeastern parts of India and the Gangetic plains during the monsoon months and the southwest coast. In contrast to this, the northwest part of India is practically dry – in fact it is a desert (Rajasthan Desert). A good part of Central India and the Deccan Plateau also suffer from serious shortages of water every year. One very rough estimate suggests that one-third of India has more water than it needs and one-third suffers from serious shortage. In the final one-third, the water is more or less sufficient, but even here in many years, because of the uneven distribution of rainfall, there is crop failure and consequent sufferings.

The best and the surest way of harnessing the water resources of India is, therefore, to store the monsoon flows in reservoirs and utilize them for stabilizing the *kharif* irrigation, and to extend irrigation in the *rabi* season to the extent that the flow and the storages permit.

In spite of the massive efforts undertaken so far, the total quantum of storage built in India till 1976 can be placed at 160 million acre-ft. The table below indicates some of the principal storages – existing and under construction – in India.

This, as compared to the average annual flow of 1500 million acre-ft is hardly adequate to have effective control and optimum utilization. One basic difficulty in this regard is the

River system/storage site		Live storage	
		million m^3	million acre-ft
	Storages of more than 2500 million m^3		
1.	Bhakra	7450	6.04
2.	Pong	6970	5.65
3.	Rihand	8980	7.28
4.	Gandhisagar	6900	5.60
5.	Hirakud	5830	4.73
6.	Nagarjunasagar	7730	6.27
7.	Pochampad	3170	2.57
8.	Ukai	7100	5.76
9.	Srisailam	5090	4.13
10.	Sharavathi	6540	5.30
11.	Koyna	2690	2.18
12.	Tungabhadra	3710	3.01
13.	Mettur	2660	2.16
14.	Balimela	2840	2.30
	Storages between 1250 m^3 and 2500 million m^3		
1.	Bhadra	1790	1.45
2.	Kadana	1220	0.99
3.	Rana Pratapsagar	1590	1.29
4.	Mahi Bajajsagar	2010	1.63
5.	Hidkal (Ghataprabha)	1420	1.15
6.	Krishnarajasagar	1250	1.01
7.	Jayakwadi	2070	1.68
8.	Bhima	1700	1.38
9.	Tawa	2100	1.70
10.	Iddiki	1470	1.19
11.	Maithon	1360	1.10
12.	Panchet	1330	1.08
13.	Ram Ganga	2210	1.79

lack of suitable storage sites on the Ganges and the Brahmaputra. Also the southwestern coast has very limited storage potential. It is in this context that the need for long distance water transfer, including storage at other sites, becomes relevant and important to India's future proposals for water resources development.

Long distance mass transfer of water has been practiced in India for over 5 centuries. The Western Jamuna Canal and the Agra Canal, built in Mughal times, are examples where water was carried from the Himalayas to the distant parts of Punjab, Uttar Pradesh and Rajasthan. In the last century, the waters of west-flowing rivers in Kerala in the southwestern part of our country were diverted to the eastern dry plateau. In the middle of the 19th century, large-scale canal construction was undertaken from the Ganges, the Godavari and the Krishna to transport water across numerous streams and valleys for extending irrigation benefits.

Compared to the present activity of mass transfer of water, these attempts of the previous century pale into insignificance. To quote a few examples – the Rajasthan Canal Project, which is now under construction in northeast India, will provide irrigation for more than 3 million acres. The water is transported all the way from the Himalayas to the deserts of Rajasthan through a series of storages, diversion barrages and canal systems. Lands which were once barren, infertile and sand-dunes are now humming with activity, with green pastures, verdent forests and teeming populations. One has only to visit some of the villages and towns on the banks of these canals to witness the transformation that has taken place in this part of the world with the mass transfer of water from the Himalayas. One of the earliest projects was the Ganga Canal in the Bikaner district of Rajasthan built in 1927. It irrigates over 600,000 acres and has transformed what was once a desert into a prosperous district.

The Rajasthan Canal Project, estimated to cost over 5000 million rupees, comprises the construction of a huge multipurpose project across the Beas River at Pong, a barrage at Harike and a grand canal system. The feeder canal from the barrage up to Rajasthan border runs for 178 km and carries a discharge of 18,500 cusecs. The Rajasthan Canal is 469 km long with numerous branches and distributaries.

The Pong Dam is now complete and more than 4 million acre-ft of water is stored behind it. The Rajasthan Canal System is also more than half complete, with work going on at full speed on the rest of the system. Today the investment on this project is of the order of 300 million rupees and each year an area of 200,000 ha is being added to the irrigation potential.

Another notable achievement of the present times is the Sarda Sahayak Project in Uttar Pradesh in north India. This project envisages transport of water from the Ghagra River to the plains of the Ganges over an area of 6 million ha. The project comprises construction of two barrages, a link channel to transport 17,000 cusecs and a feeder canal 260 km long to deliver supplies to various existing channels. It also envisages remodelling and improvement of the existing canal system to provide adequate and efficient water conveyance. The cost of the project is estimated at over 2000 million rupees and on completion will provide irrigation to 4 million acres. To date, 80% of the work is complete and already more than a million acres are receiving irrigation benefits.

Another equally important major project is the Ram Ganga in Uttar Pradesh. Here again, the waters of the Ram Ganga, a tributary of the Ganges, are being stored in the Ram Ganga Dam and transported south to various districts for assured irrigation to over 1.5 million acres. This project is almost complete.

Many other big projects completed in recent years also envisage large-scale mass transfer of water. The Bhakra–Nangal, the Nagarjunasagar and the Tungabhadra are giant schemes, irrigating 4 million, 2 million and 1 million acres, respectively. Volumes can be written about these projects and their effect on the environment and people in the area benefitted by these projects.

The long distance mass transfer of water definitely has an influence on the environment of a place. Mention in this connection may be made of the improvement in salinity in the Godavari and Krishna Deltas as a result of the introduction of irrigation through large canals flowing over long distances. But for the introduction of irrigation, a good part of these deltas would still be affected by high salinity and a high incidence of malaria and other pestilence.

LONG DISTANCE TRANSFER – THE FUTURE PERSPECTIVE

Notwithstanding the massive development of irrigation that has taken place during the last 30 years since independence, the Government of India and the State Governments are now engaged in planning and investigating a large number of schemes for mass transfer of water. For, it is obvious that unless such mass transfer is carried out, there is little opportunity of providing even the basic facilities of a single crop and drinking water in most parts of the country. It is a common feature every year to hear stories of large-scale water shortages even for drinking and the subsequent mass movement of population and cattle. The planning for irrigation development has, perforce, to take into account this essential feature of the Indian topography and climate and so the schemes now under contemplation visualize large-scale mass transfer of water from one part of the country to another. Some important schemes are briefly described below.

Godavari–Krishna–Pennar link

The Godavari River, the largest river in the Indian peninsula, according to preliminary studies, has a surplus of water, whereas the Krishna and the Pennar Basins as compared to the Godavari, have more land potential than water. The Pennar, in particular, has a serious shortage of water in its basin. Schemes for transporting water from the Godavari to the Krishna and the Pennar have, therefore, been under investigation for more than 70 years now and various alternatives have been contemplated at different times. Just at about the time of Independence, a gigantic scheme was proposed by the then Madras Presidency for transporting water from the Godavari to the Krishna and from the Krishna to the Pennar up to the outskirts of Madras city. For various reasons this project did not see the light of the day at that time. Recently, however, in the sixties, an alternative of transporting nearly 10 million acre-ft of water was conceived. Some detailed investigative work has been done in this connection, but because of certain inter-state aspects involved in such a transfer of water, there has been some delay in its implementation. At present, a more limited scheme to benefit the areas within a State from the mass transfer of water are now being formulated. Transfer of water within the State from the surplus areas to the deficit areas is envisaged. The Godavari waters are for irrigation and to meet drinking and industrial needs of the steel complex at Visakhapatnam. Parts of the water will also be transported southwards to augment the flows in the Krishna and from the Krishna to the Pennar basin.

Almost the entire Pennar basin in Andhra Pradesh is drought-prone. The irrigation and drinking water requirements of this area can be met – topography permits it – by the diversion of the Godavari and Krishna waters to this area. The proposals now under consideration envisage such a diversion, but the finalization of the scheme will depend on the final award of the Godavari Tribunal which has yet to pronounce its judgement. In the case of the Krishna, the Krishna Tribunal has already given its verdict and therefore there should be no difficulty in taking some of the surplus flows in the Krishna to the deficit areas in the Pennar basin by cutting across the ridge and by extending canals. Detailed investigations are now in progress for this scheme.

The Narmada High Level Canal

Another major proposal for mass transfer of water is the Narmada High Level Canal Scheme of Gujarat State. In this scheme, the construction of a high dam at Navagaun in Gujarat is proposed. Leading from this dam a high level canal will be built crossing numerous rivers and streams and extending into North Gujarat and the desert areas of Kutch. Parts of Rajasthan can also be benefitted by this canal. The scheme, as formulated by the Gujarat Government, envisages a canal of 15,000 cusecs capacity, 600 miles long and benefitting 5.7 million acres. In this case also there is an inter-state dispute between the Gujarat State and the other basin States of Madhya Pradesh and Maharashtra. A Tribunal has been in session examining this matter and a decision is expected soon. The implementation of the scheme has, therefore, to await the verdict of this Tribunal.

The Scheme formulated by the Gujarat Government will completely change the face of the scarcity areas of North Gujarat and the saline areas of the Rann of Kutch. This is an instance where the environment will be completely transformed with the introduction of water by mass transfer from the Narmada river over a long distance.

Preliminary estimates have placed the cost of the scheme at over 6000 million rupees. The correct figure can be worked out only at the time of implementing the scheme, as the cost of materials and labour have been fluctuating from time to time.

West-flowing rivers

Another major possibility that exists in India for large-scale mass transfer of water from one area to another, is the diversion of the west-flowing rivers to the east to provide irrigation facilities in the drought-prone areas of Andhra Pradesh, Maharashtra, Karnataka and Tamil Nadu.

As indicated earlier, the west coast has a surfeit of rainfall and river flow. The land area that can be benefitted from this river flow is limited and even assuming that the entire land in this stretch will be provided with high intensity of irrigation, there is, according to indications, surplus water available. What is required is a careful assessment, field investigations and finalization of schemes for the conservation and transfer of the surplus waters to the eastern side of the Western Ghats to meet the irrigation and drinking water requirements there. It is in this connection that studies are in progress at various levels. According to some rough indications, more than 200 million acre-ft of water are now going waste into the Arabian Sea and even if a small part of it is harnessed this way, it will provide tremendous relief to the drought areas in Tamil Nadu and other States.

The Ganges lift schemes

While the plains north of the Ganges are blessed with numerous tributaries and good groundwater aquifers, the southern portion of the Ganges basin, south of the Ganges, consists mostly of broken land mass, criss-crossed by numerous streams which are not often perennial. There is great scarcity and shortage of water and parts of this area are often prone to drought. A number of dams, mostly medium and small, have been built, but the total effect of all these on the people and the land has not been appreciable. Schemes are, therefore, being formulated to provide irrigation facilities in the southern half of the Ganges basin in a big way, by lifting water from the Ganges and the Yamuna and providing direct irrigation in as large a part of the area as possible. It is also proposed to store a part of the lifted waters where feasible and then extend irrigation. This appears to be the only way of solving the problem of this drought-prone area. While most of the earlier schemes were confined either to small storages or to low lifts, the present thinking is on high head, large-scale lifting of water and storage. It also envisages interlinking of numerous tributaries of the Ganges through canal systems. Preliminary studies have indicated that there are very good possibilities for such lift schemes and detailed investigations are being organized for this purpose.

Brahmaputra–Ganges Link

The Brahmaputra carries very large flows not only during the monsoon season but also in fair weather. Recurring floods of this river cause great loss of life and property both in India and in Bangladesh. Therefore, the possibilities of control and development of the Brahmaputra are currently under consideration. The Brahmaputra rises two months ahead of the Ganges and a Brahmaputra–Ganges Link, supplemented by storages, could enable integrated development, and would enable flood control, power generation and optimum utilization of the water resources of the lower Ganges–Brahmaputra region for the benefit of the two countries. It is hoped that studies on these possibilities may start in the near future in cooperation with Bangladesh.

The Rajasthan Desert

As already mentioned, part of the Rajasthan Desert is deriving a great benefit by the mass transfer of water through the Rajasthan Canal, the Gang Canal and other irrigation projects which are already complete or nearing completion. However, schemes are still under formulation in respect of quite a few other major projects. Mention may be made in this connection, of the storage scheme on the Yamuna whereby the flood waters of the Yamuna are stored at a place near Kishau and the water transported to the deserts of Haryana and Rajasthan. Proposals are also under consideration for building storages on the tributaries. A preliminary paper scheme has also been formulated for transporting some water from the Chambal River to the northwest parts of Rajasthan through both lift and long distance carrier system. These and many other proposals are still being investigated.

In this connection, mention may be made that the Central Government, realizing the importance of mass transfer of water for development and a better environment, have recently constituted a new Investigation Unit in the Central Water Commission. A full-time Chief Engineer has been appointed with necessary field staff and technical supporting staff at headquarters to prepare a number of feasibility studies for long distance mass transfer of water. In fact, India can take a justifiable pride in having made an earnest effort in national water planning and in the coming decade or two it is hoped that a number of these mass transfer schemes will fructify and change the human environment in many parts of the country.

CONCLUSION

With the advance in science and technology the world over and the refinement in the techniques of high head pumping and tunnelling, the dream of taking water over long distance for the benefit of man no longer remains a vision. Large-scale mass transfer of water has become a reality. Keeping in view that nearly one-third of the country is drought-prone, such transfer of water will definitely usher in a new era of better environment for the people in these drought-prone areas and also lead to a change in the ecology of these places. It is a fact that with the introduction of irrigation, the vegetation, the fauna and the flora change, thereby altering the ecology of the place. Such improvements have added advantages of a chain reaction in many spheres which lead to a more prosperous life for the people of the area. The economics of long distance transfer of water has to be viewed in this context and also the appreciable savings in millions of rupees that are spent at present on relief in drought affected areas.

The mass transfer of water is one major answer to the two faces of water – floods and droughts – in India and in the rest of the world. An earnest endeavour in understanding the implications and improvements in the technology of such transfer would be of great benefit to the people of India and to mankind as a whole.

PART II: Problems of Interregional Water Transfers

Water Supply & Management, Vol. 2, pp.127-136. 0364-7714/78/0601-0127$02.00/0

Economic Issues Related to Large-scale Water Transfers in the USA

CHARLES W. HOWE*

I. EXISTING AND PROPOSED TRANSFERS

The distribution of precipitation in the United States is quite uneven. In very crude terms, the eastern half of the country (east of 100° west latitude) is well watered while the western half is dry. It is not surprising then, that water quality problems predominate in the East while water quantity problems predominate in the West. The greatest interest in interregional transfers has therefore been in the West. However, large metropolitan concentrations of population often demand more water than can be found in their immediate drainage basins, so transfers have been undertaken to large cities even in the East.

New York City developed one of the earliest systems, starting with staged development of the Croton River, a distance averaging 250 km, over the period 1842–1904. The Catskill system, averaging 400 km in distance, was built over the period 1915–1924. Together, these two systems provide about 1.21×10^9 m^3/yr. In 1936, development of the Delaware River, which is shared with the States of Delaware and Pennsylvania, was begun, culminating in a system of large reservoirs and aqueducts with a safe yield of 1.3×10^9 m^3/yr.

Conflicts of interest accompanied the Delaware development. The State of Delaware tried to prevent New York City from transferring water, even though the Delaware River rises in New York State, then flowing into the State of Delaware. A decree of the Supreme Court permitting the city to divert water while requiring the city to meet minimum releases from its reservoir system was required to settle the argument.

The prolonged drought of 1961–1966 caused the estimated "safe yield" of the entire New York system to be reduced from 2.46×10^9 m^3/yr to 2.0×10^9 m^3/yr. An interesting feature of the New York City system is that the Hudson River which flows through the City has not been used for water supply, even though the Croton and Catskill units are in the upper Hudson River drainage. Economists have argued that water from the Hudson could have been developed at a fraction of the cost of the Delaware system. The City's Department of Water Supply has counter-argued that pure sources of supply justified the additional cost.

In the West, the State of California exhibited the earliest large interregional transfer and has recently completed the largest one. The City of Los Angeles built the Los Angeles Aqueduct in 1913 to bring water from Owens Valley on the eastern side of the Sierra Nevada Mountains, a distance of 300 km. This aqueduct was extended on to Mono Lake, a distance of 500 km for a total yield of 580×10^6 m^3/yr. Severe controversy surrounded the Owens Valley development, for the valley residents didn't want to give up the agriculture based on the water. The City finally bought the lands of the valley, but some parties continued to resist the building of the aqueduct.

---

* Professor of Economics, University of Colorado (Visiting Professor, University of Minnesota, 1977–1978).

Continued growth of the Los Angeles area led in 1928 to the construction of the 400 km Colorado River aqueduct to tap California's share of that river. This aqueduct currently delivers 1.5×10^9 m^3/yr to the south coastal area. Some of this water is used to recharge coastal aquifers from which much pumping takes place.

The US Bureau of Reclamation in 1935 started the Central Valley Project (California) to capture and transfer mountain waters from Northern California to points along the San Joaquin Valley with distances up to 600 km, delivering 3.4×10^9 m^3/yr. This system has been supplemented by the State Water Project along similar lines, capturing Feather River water in the north and transporting, in total, 5.2×10^9 m^3/yr to the San Joaquin Valley and the Los Angeles area. The 2.5×10^9 m^3/yr going to Los Angeles travels as much as 800 km and must be lifted 610 m over a range of mountains.

The severe drought affecting Northern California in 1975–1977 has affected the yield of this system severely, and Los Angeles received no water from the State Water Project during part of the summer of 1977. The State Water Project has been severely criticized by environmental interests for damming the scenic Feather River and by others for the high cost of the project.

Smaller transfers are found in other parts of the West, but among the larger and more important is the Colorado–Big Thompson project which transfers water about 80 km across the Rocky Mountains to eastern Colorado for irrigation and municipal use. While this system transfers only 370×10^6 m^3/yr, it has provided a vital supply for a rapidly growing region of Colorado. The institutional arrangements for distributing the water and for allowing transferability of the water among uses, are nearly unique in the US and will be described in Section II.

Discussions of new large-scale transfers for the western United States reached a peak in 1967 or 1968. After that time, interest waned quickly, first because of the strong objections of the potential exporting basins, and later because of rapidly rising costs. Since the oil embargo of 1973 and the severe drought of 1976, interest has been somewhat revived. Some particular regional problems have also led to renewed interest.

The most actively debated interregional transfers during the mid–1960s were several plans for Columbia River Basin transfers. These transfers were designed to carry from 3×10^9 m^3/yr to 18×10^9 m^3/yr. Several of the plans called for taking water from the Lower Columbia River which had the effect of substituting higher pumping costs for greater instream opportunity costs of the water. Other plans called for taking the water from the tributary Snake River at higher elevations, saving on pumping costs but incurring greater foregone uses downstream; primarily foregone hydroelectric power. During the past two summers, it would have proven impossible to export water from the Snake River because of extreme drought in the Northwest (see Fig. 1).

The only active proposal involves the possibility of importing water into the high plains region of western Texas and eastern New Mexico, a region where a highly productive irrigated agriculture has been developed from the use of groundwater. The entire regional economy is dependent on the current high yields of agricultural commodities, but the groundwater is being exhausted. Only in recent years has there been any effective attempt to control the use of these non-renewable groundwaters – much too late in terms of an optimum strategy from a national point of view. This kind of water import situation has been referred to as a "rescue operation" because the region itself cannot afford the large water transfers necessary to maintain its economic base.

A potential source for the 6×10^9 m^3/yr which would replace current consumptive uses of groundwater, is the lower Mississippi River, probably by pumping the water up the Red

River Valley, involving a distance of some 1300 km and an attitude difference of 1200 m. Since no power recovery is possible, this route currently has prohibitive energy costs associated with it. A second route currently being discussed involves taking water from the State of Arkansas and transporting it through a system which the intervening state of Oklahoma would

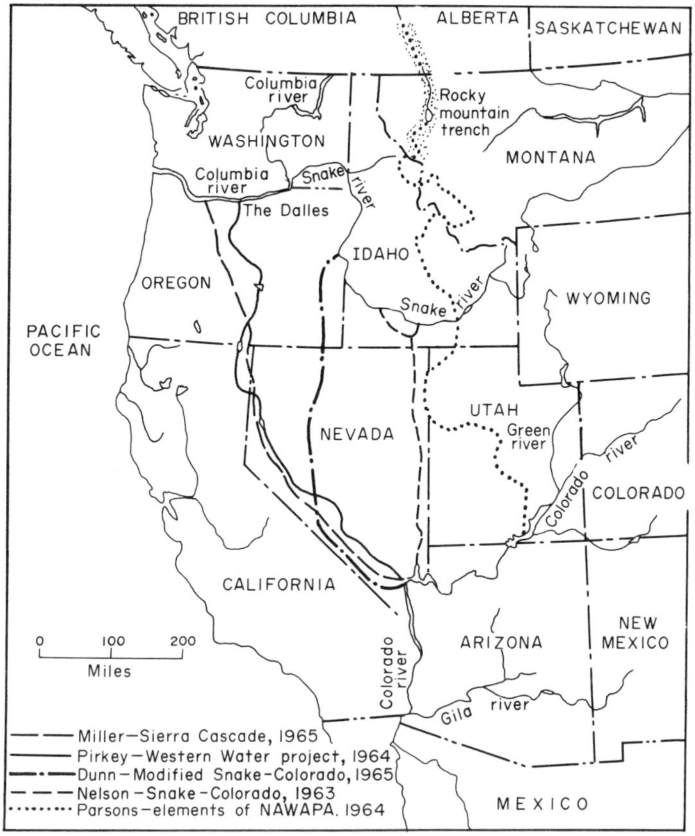

Fig. 1. Five interbasin water transfer projects.

like to develop for similar purposes. The incremental distance in the length of the canal system could be as little as 500 km and the altitudes difference would be much less than the other route.

An important factor in the entire North American water transfer picture is the opposition of the areas of origin to proposed water transfers. Canada took a strong position (e.g. statement of John H. Turner, Parliamentary Secretary to the Ministry of Northern Affairs in the mid–1960s) that Canadian water was for Canadian development, and only after the most careful studies of potential Canadian uses would Canada consider exporting water to the US. The states of the Northwest have solidly opposed exports from the Columbia River Basin, and their united political power was sufficient to prevent the various river basin commissions and even the National Water Commission (formed to study US water policy and problems) from considering or studying interregional transfers. Senator Jackson of Washington has stated:

> The people of the Northwest deeply believe that before any other region asks for a study of the diversion of the Columbia River, such region must first establish that it actually needs additional

> water ... What for? ... Can sufficient water be secured through conservation and reuse? ... How will the economy of the Northwest be affected if large quantities of water are taken away?

This opposition emphasizes not only the need for the studies called for by Senator Jackson, but the need to consider new institutional arrangements within the United States for managing these large transfers if they occur and for providing compensation to the areas of origin. Primary jurisdiction over water is held by the States. Distribution of the waters of interstate rivers has been decided by interstate compact (treaty) in the arid regions and generally remains undecided in the water plentiful regions. The River Basin Commissions which exist to coordinate planning *within* a major basin are expressly forbidden to consider transfers from outside their drainage areas. Only the Federal Bureau of Reclamation is in a position to consider transfers and to put together compensating programs for the areas of origin. However, compensation is limited to the construction of more water projects. This has proved to be a costly, inefficient way to provide regional compensation.

II. ECONOMIC BENEFIT ISSUES RELATED TO TRANSFERS

A. *Low benefits in agriculture*

The regions of the US where calls for imports of water are most frequently heard are characterized by arid climates and irrigated agriculture as the largest consumptive use. Examples would be the Lower Colorado River (especially Arizona and California) and the High Plains of western Texas and eastern New Mexico. While these areas are highly productive in physical terms and are partly devoted to speciality crops of high value, they also contain vast areas of low value crops, especially forage crops and low value feed grains. Net income per hectare often does not exceed \$100 per yr, i.e. as little as \$0.008 per m^3 of water applied. While multiplier effects might raise this value to \$0.016 per m^3 from the region's point of view, it is not sufficient to justify transfer costs of at least twice that amount.

The relevant comparison for evaluating transfers is between the lowest values in agriculture and the unit transfer costs, since water can almost always be transferred from the lower valued uses to speciality crops or even to industries and city use if those demands grow. An excellent reference, describing the economic structure of the arid Southwest and the value of water to agriculture there, is Kelso *et al.* (1973) who studied the effects of the falling groundwater table on the economy of the State of Arizona. They projected, using linear programming models, the likely reductions in cropped acres, consumptive water use, gross value of farm output and net farm income caused by increased pumping costs. Their findings nicely illustrate the low marginal value of water in agriculture in terms of regional income. Some relevant data are given in Table 1 below.

B. *Agricultural displacement effects*

During the period from 1950 to 1965 or somewhat later, US agriculture was faced with a continuing problem of surplus production. World markets were poorly organized, foreign aid for food purchase was not well established until later in that period, and domestic US demand was income and price inelastic. During this period, agricultural technology was rapidly advancing, raising productivity and lowering costs. As a result, farm prices were kept low and large surpluses accumulated in government hands. Under such conditions, in a market

Table 1. Projected declines in Arizona agriculture due to rising groundwater costs: 1966–2015

Cropped area (1000 ha)	119	28%
Water use (10^6 m^3)	1684	29%
Gross value farm output (millions of dollars)	37	13%
Net farm income (millions of dollars)	17	15%
Direct income loss per m^3 (dollars per m^3)	0.006	
Direct plus indirect income loss	0.007	

economy, the opening of new acreage either depresses prices further, driving existing farms out of business or causes government surpluses to increase or both.

The relationship of these observations to many proposed interbasin transfers is that many transfers are designed to provide irrigation water. If the newly irrigated lands are to be profitable to the farmers when crops nationally are already in surplus, it may be necessary (as it was in the US) to charge much less for imported water than its true cost, for when the new lands come into production, they will lower prices. This can have the effect of driving equivalent acreage out of production elsewhere. These effects for the US during 1944–1964 are documented in Howe and Easter (1971). Planners must be aware of the possibility of these problems in the future, even though world market conditions have changed.

C. *Efficiency of use of transferred waters*

Transfers are usually planned into growing regions whose economies are changing rapidly. In market economies, it may be difficult to predict the future structure of the region's economy, e.g. which industries will be there, what the urban population will be, etc. It is therefore important that the institutional arrangements made for the allocation of the imported water allow for changing priorities and water demand patterns over time.

The arrangements developed by a large water administration district in Colorado have been particularly innovative and efficient in this respect. The Northern Colorado Water Conservancy District was established to develop and distribute water in a 2300 square mile area of northeastern Colorado. While some local river flows were available for distribution, most of the water was to be provided by a new federal storage and diversion project, named the Colorado–Big Thompson project, which diverted water from the western slopes of the Rocky Mountains to the eastern slopes and plains. The amount of water handled annually is about 380×10^6 m^3.

Water allocations were originally made to landowners, municipalities, and industries in the District, irrigation being by far the largest user. The "shares" so distributed are freely saleable among parties located within the District, so permanent sales of water from less productive to more productive uses can take place.

Seasonal water "rentals" also take place. If a farmer finds that his allotment for the year is more than his planned crops or livestock will require or if high prices offered for the seasonable transfer of water make it attractive to reduce his applications, he can advertise through the District office that some of his water is available for sale for this year at whatever price he cares to ask. Farmers seeking additional water can then bargain with sellers of water, and a very smooth market process has developed. Seasonal rental prices of water sometimes reach \$25 per 10^3 m^3.

As a result of this ready market for water, water use is carefully planned by the farmers. Since there is no danger of losing one's permanent water rights by a sale of part of one's annual allotment, farmers prefer to sell the water if its value rises above the return they can obtain on their own farm. Economic efficiency of water use is very high.

D. Benefits along the transfer route

While most transfers are initially thought of in terms of an area of origin and a distant area of destination, it may turn out that investigations will uncover potentially beneficial uses along the transfer route which can be served at low marginal cost. It may also be politically necessary or advantageous to include some developments en route to secure the backing of the regions through which the transfer will pass.

In the US, some of the Columbia River–Colorado River transfers would have passed through semi-arid farmland where supplemental irrigation would have increased productivity. Whether benefits would have offset incremental system capital and operating costs was not investigated. Industrial projects may beneficially be expanded into multiple purpose projects, as with the Shashe Project in Botswana, which captures and transfers water about 80 km to a copper–nickel smelter complex, passing through arid areas in which severe village water supply problems exist. These problems might have been dealt with at low marginal cost by designing several small pipelines off the main trunk line.

E. Secondary benefits

In predicting the impacts which large projects will have on regional and national economies, the question of "secondary benefits" always arises. The most usual definition of secondary benefits is "benefits legitimately countable from a national viewpoint accruing to parties other than direct project beneficiaries". Two features of this definition should be emphasized for purposes of correct economic analysis: (1) that the benefits should be net additions from a national viewpoint and not simply a transfer from one region to another; (2) that they accrue initially to parties other than direct project beneficiaries. The first feature tells us, for example, that if a processing industry shifts to a riverside location because the river has become navigable, only the cost-savings it experiences can be counted as economic benefits, not the net value of its total output. The second feature reminds us that we must not double-count benefits which initially accrue to direct beneficiaries and are later passed on to others through the market or because of central direction to do so. Thus, if the cost of supplying water to Industry A is reduced by a transfer and if, as a result of this cost reduction, the unit price of the industry's output is reduced, one must not count *both* the initial cost reduction and the lowered price to Industry A's customers as benefits.*

The reason for raising this issue in our present discussions of interregional transfers is that the secondary benefit concept has been greatly abused by the water resource agencies in the United States as a way of overstating water project benefits. Practices have included

* The details of the final incidence of the benefits are difficult to determine. Of course, allowance must be made for possible changes in Industry A's output rate made profitable by the cost change.

counting the gross output (sales) of project related enterprises (such as farm suppliers) and counting the outputs of existing industries which simply shift location because of minor cost advantages associated with being close to the water project.

However, interregional transfers are likely to be large projects relative to the size of the regional economy, so secondary benefits and costs must be analyzed, preferably from both national and regional points of view.

III. ECONOMIC COST ISSUES RELATED TO TRANSFERS

A. *Energy intensity and energy recovery*

Preliminary design studies of several Columbia River–Colorado River transfer systems and several routes for the transfers to West Texas have shown clearly that the amount of pumping which must be done and the amount of energy recovery which is possible are crucially important. From the Lower Columbia Basin, a lift of at least 1800 m would be necessary, but the water would be delivered to the Colorado River at an altitude of about 600 m so that a substantial amount of energy recovery through electric generation would be possible. Transfers from the Snake River would originate at a higher altitude reducing pumping requirements but increasing the opportunity cost of the water and reducing the reliability of the supply. Transfers from the Mississippi to West Texas involve the large attitude differences noted earlier with no possibility of power recovery. With energy costs at current levels, the latter transfer is grossly infeasible from economic and financial viewpoints.

B. *Water opportunity costs and other externalities*

These costs are frequently ignored in US water planning, largely because States' legal claims to water or the allocations under interstate compacts (treaties) are at variance with the criterion of economic efficiency. The major forms of opportunity cost are foregone irrigation uses, foregone power generation and deterioration of water quality because of reduced dilution. Reduced esthetic values and reduced sport fishing have occurred as a result of transmountain transfers to the Denver metropolitan area.

Increases in salinity concentrations reduce agricultural yields, impose additional costs on municipal and industrial systems, and at times severely impact coastal zone fisheries. The Gulf (of Mexico) Coast of the US has suffered reductions in important shrimp and oyster catches, and the west coast of Mexico has suffered major reductions in its shrimp fishery, both because of reduced fresh water flow.

IV. THE TIMING OF LARGE INTERREGIONAL TRANSFERS

Scale economies in all water transfer technologies and the low unit value of water imply that interregional transfers must be large to be economically feasible. Large increments to regional water supplies by definition imply that timing of the project is very important. Even with water demands in the receiving region growing, premature construction will mean unused capacity for long periods of time, while deferring construction to allow demand to grow closer to the transfer's designed capacity implies either the interim use of costly short term supplies or a delay in regional growth.

Three large transfers in North America, the Plan Hidraulico de Noroeste for the Costa de Hermosillo in northwest Mexico, the Mississippi–West Texas transfer and the Central Arizona Project in the US were intended as "rescue operations" – the provision of water to replace exhausted groundwater. In such a case, timing is crucial from the economic and possibly from a physical viewpoint.

In the Costa de Hermosillo, highly productive commercial agriculture had been established on water pumped from a large coastal aquifer. Pumping exceeds recharge by a wide margin and salt water intrusion from the Pacific Ocean is proceeding at a rate of over 1 km/yr. Several questions were raised:

(1) From a purely economic viewpoint, when (if at all) should an alternative supply be developed?

(2) Can the salt water intrusion be reversed in the future through artificial recharge and/or reduced pumping?

(3) What economic cost is worth incurring to avoid the possibility of irreversible loss of large parts of the aquifer?

Regarding the first, Ronald G. Cummings (1974) analyzed the economics of a large transfer of water up the West Coast to the Costa de Hermosillo. Through a large programming model linked to a digital model of the aquifer, he was able to show that very large quantities of water could still be economically mined from the aquifer, with the optimal rate of pumping gradually approaching the recharge rate over a 36-year period. The shadow price of water in the aquifer, giving its real scarcity value, was shown to equal the estimated unit cost of imported water only 29 years from now. Thus, from a purely economic point of view, construction should be delayed *many* years, saving many millions of dollars in terms of the present value of project costs. The results of one model run are given in Table 2 below.

Somewhat similar conditions are faced in the central part of Arizona, a rapidly growing region between the cities of Phoenix and Tucson. In this area irrigated agriculture has been quite important historically, but the growth is in light industry, commerce and services for the retirement communities. Groundwater is the main water source and, largely because of agricultural uses, the water table is falling from 3 to 6 ft/yr. In some places, pumping depths are over 600 ft (183 m). The aquifers are very deep and vast quantities of water remain available, but costs are increasing and surface subsidence has become a problem.

The Central Arizona Project, while not "interregional" is a major pumping project to lift 1.5×10^9 m^3/yr of water from the Colorado River to replace part of the groundwater being used by agriculture and municipalities. Given the continued availability of groundwater and a rather smooth market process of transferring groundwater stocks from agriculture to municipalities as the urban areas grow, the question of the optimum time of construction arises. Farmers seem unwilling to pay more than about \$0.008 per m^3 since they can pump water at that cost, and cities are reluctant to pay the price of \$0.04 per m^3 which has been proposed since they, too, can pump water from the lands into which they are expanding. Nonetheless, there is a long-term problem, the solution to which *should* have involved estimating optimum timing of the transfer.

V. FINANCING INTERREGIONAL TRANSFERS

The main point to be made here is that inefficient large-scale water projects are much less likely to be undertaken if public financial policy calls for the direct and secondary beneficiaries to pay a major portion of the construction, operating and maintenance costs. With the

Table 2. Optimum use of groundwater: Costa de Hermosillo

Year	Annual rate of pumping (million m^3)	Groundwater storage at the beginning of of year (million m^3)	Increase in storage attributable to pump relocation (million m^3)	Shadow value of water not discounted) (dollars/m^3)	Increase in saltwater intrusion (km)
1	1,219.1	22,253.0	1,989.6	0.0008	0.96
2	1,219.1	23,023.6	795.3	0.0035	0.96
3	1,219.1	22,234.0	828.1	0.0038	0.96
4	1,219.1	21,412.0	829.4	0.0042	0.96
5	1,219.1	20,588.7	829.5	0.0046	0.96
6	1,219.1	19,765.3	829.5	0.0051	0.96
7	1,219.1	18,941.9	829.5	0.0054	0.96
8	1,219.1	18,118.5	829.5	0.0060	0.96
9	1,206.3	17,295.1	143.3	0.0067	0.95
10	1,206.3	17,834.8		0.0074	0.95
11	1,206.3	16,941.7		0.0080	0.95
12	1,206.3	16,048.6		0.0089	1.7
13	1,206.3	15,091.2		0.0096	1.7
14	1,218.6	13,976.3		0.0109	1.7
15	1,202.2	12,806.5		0.0118	1.7
16	1,126.5	11,638.3		0.0122	1.4
17	1,048.7	10,546.3		0.0124	1.4
18	978.5	9,552.7		0.0134	1.3
19	915.3	8,656.0		0.0138	1.3
20	865.1	7,848.5		0.0141	1.0
21	796.5	7,115.5		0.0156	0.9
22	756.7	6,471.1		0.0173	0.9
23	603.5	5,890.4		0.0175	0.5
24	555.3	5,480.2		0.0178	0.4
25	552.5	5,164.1		0.0184	0.4
26	527.0	4,876.2		0.0188	0.4
27	524.6	4,621.5		0.0192	0.4
28	512.5	4,378.2		0.0203	0.3
29	510.3	4,150.0		0.0224	0.3
30	508.1	3,928.0		0.0246	0.3
31	506.0	3,709.9		0.0272	0.3
32	503.9	3,495.1		0.0296	0.3
33	501.8	3,283.1		0.0320	0.3
34	500.0	3,074.1		0.0360	0.3
35	497.7	2,868.0		0.0360	0.3
36	350.0	2,644.5		0.0400	0.3

Source: Cummings, 1974, p. 98.

early US transfers like the New York City system, financing was completely done by the water utility itself through bonds which were paid off through volume charges to water users. As the federal government has come to dominate the planning and funding of large water projects, the degree of subsidy has grown greatly. This has served to make inefficient projects look desirable to local interest groups who then attempt to rally political support for their favorite projects. A solid policy of full cost recovery on transfer projects will be of great assistance in guarding against inefficient projects.

REFERENCES

Cummings, Ronald G. *Interbasin Water Transfers: A Case Study in Mexico,* Johns Hopkins Press, Baltimore, MD, (1974).

Howe, Charles W. and Easter, K. William. *Interbasin Transfers of Water: Economic Issues and Impacts,* Johns Hopkins Press, Baltimore, MD, (1971).

Kelso, Maurice M. *et al. Water Supplies and Economic Growth in an Arid Environment,* University of Arizona Press, Tucson, (1973).

Water Supply & Management, Vol. 2, pp.137-145. 0364-7714/78/0601-0137$02.00/0

Some Theoretical and Measurement Issues in Economic Assessment of Interbasin Water Transfers

ANTHONY C. FISHER*

1. INTRODUCTION

With the tremendous increases – of an order of magnitude or more – in size and cost of recently proposed interbasin water transfers (IWT's) over those of existing projects, careful consideration of their economics takes on new importance. In this paper I offer some critical remarks about concepts and measurement techniques in assessing the costs and benefits of IWT's. I say critical, because I think that is what is needed, but I hope to provide some constructive suggestions as well. Also, I want to acknowledge, right at the outset, that I speak as a relative outsider, one not familiar with much of the work in the field. This has obvious drawbacks, but perhaps, in view of the ample expertise represented at this Conference, it may be useful to hear a fresh voice.

The remarks will fall into three categories: methods of measuring conventional economic costs and benefits, introduction of environmental effects, and special problems posed by the very long-lasting and uncertain consequences – including those to the environment – of IWT projects. Not coincidentally, these categories are listed, and will be treated in the sections to follow, in order of decreasing specificity. That is, I hope to be specific and constructive about methods of measuring conventional benefits and costs. About the environment, I can be specific with respect to the problems but not very helpful with respect to solutions (though I do have one or two ideas), and about long-run uncertainty I fear I can indicate only in a rather vague way the nature of the problems this poses for economic assessment, and suggest some qualitative policy implications.

The current "best practice technology" for assessing the impact of an IWT on a region's economy (presumably positive for a region gaining water or transmitting it, negative for one losing, or competitive with the gaining region) is input–output (I–O) analysis. The critical part of my remarks in the next section will be to the effect that I–O, especially of the required regional variety, is not entirely adequate to address the concerns of decision makers about project impacts. It does represent an advance over a number of alternative, simpler methods of regional impact analysis, as I shall indicate. But, more constructively, I shall propose the use of an econometric modeling technique that can take account – as I–O does not – of both changes in the structure of the impacted region's economy, and the time periods required for these changes to work themselves out.

The environmental problem is simply that an IWT is virtually certain to have an impact – quite possibly adverse – on the environment that is not reflected even in the most sophisticated econometric analysis. In Section 3 below, I say a bit more about the nature of the impact and indicate how it can, in principle, be incorporated into the benefit–cost analysis. Prospects for achieving a common metric – say money units – are not especially

* Professor, University of California, Berkeley.

encouraging. But the notion of *dominance*, described in Section 3, may offer a way, even without this, to comprehensively evaluate an IWT in comparison to some alternative for providing water.

The third set of problems, involving the very long time spans over which the effects of an IWT may be felt, and uncertainty about the nature of these effects, is still less tractable to conventional benefit–cost analysis. Long time spans raise questions about rates of discount, and transfers of resource endowments between generations. And there is no one accepted method for handling uncertainty, even in the short run. What I shall indicate, though, in Section 4, is that the interaction between uncertainty and irreversibility (of a project's effects) does have some rather sharp *qualitative* implications for policy. If it is known that the environmental effects of a project are irreversible, i.e. cannot be undone, except perhaps at prohibitive cost, and if it is possible over time to acquire information about the costs and benefits of the project, and its (reversible) alternatives, then there is some presumption in favor of deferring the project. This proposition is demonstrated with the aid of an example in Section 4. Unfortunately it is difficult, in the present state of our knowledge, to make any *quantitative* assessment of the "option value" of deferring.

2. MEASUREMENT OF REGIONAL ECONOMIC IMPACTS

Let us begin the discussion of measurement techniques by restating the basic benefit–cost relationships for an IWT, as presented in the important work of Howe and Easter.[1]

$$(DB_M + SB_M) + (DB_T + SB_T) > (DC_X + SC_X) + SC_C + TC \quad (1)$$

and

$$TC + [(DC_X + SC_X) - (DB_T + SB_T)] < TC_A, \quad (2)$$

where DB is the direct benefit from the water, DC the direct cost (of foregone water), SB and SC are secondary benefits and costs (to be described below), and TC is the cost of the physical transfer system. The subscripts are M = region importing water, X = region exporting, T = region through which water is transferred, and C = region whose output is competitive with M.

Inequality (1) then states that the direct and secondary benefits, to importing and transfer regions, must be greater than the direct and secondary costs, to exporting and competitive regions, plus the cost of the transfer facilities. Inequality (2) states that the cost of the transfer must be less than the cost of the best alternative, TC_A, for providing the water. All costs and benefits can be considered in present value terms (i.e. each cost or benefit term represents, where appropriate, a discounted sum). If an IWT meets both conditions (1) and (2), it is said to be economically *efficient*.

I propose to use these relationships as a framework for discussion of some specific measurement issues. In the remainder of this section I consider the measurement of the conventional economic direct and secondary benefits and costs. As noted in the introduction, the most advanced method, used in a number of studies described by Howe and Easter and also in their own work, is regional input–output (I–O) analysis. Below I briefly survey a range of alternative methods, and indicate the advantages and disadvantages of regional I–O.* Then I

* For a much more complete review of methods of regional impact analysis, see Isard.[2]

propose still another alternative, a form of regional econometric analysis, that I feel holds the promise of avoiding the difficulties associated with the earlier methods.

Regional I–O and other methods of impact analysis

Typically, analyses of the economic impact on a region of some proposed policy or resource development project employ some variant of one of the following methods: projection of past trends, economic base multiplier analysis, or regional I–O.

Simple projection, or extrapolation of past trends of such economic variables as output and employment by sector, or of demographic variables such as the school-age population, clearly are not adequate to measure the impact of a major new development on the region experiencing it. This is particularly true if, as in the case of the newly proposed IWT's, the development is quite large relative to the current economic base. In this case we can be fairly certain that past trends will in fact be modified in some way.

Economic base multiplier methods offer some improvement over simple extrapolation. The multiplier methods divide economic activity in a region into two types: basic and non-basic. Basic activity produces output for export, and non-basic other goods and services. Account is taken of the proposed development by specifying, exogenously, a new level of basic employment. This might mean, for example, employment in agriculture in an area irrigated by water from an IWT. Total employment (basic plus non-basic) and population are then forecast on the basis of multipliers, the ratio of total to basic employment, for employment, and the ratio of population to basic employment for population. The problem, however, is that the multipliers are derived from the current level and composition of employment in the region. For the forecasts to be accurate, the multipliers must remain constant, and there is no reason to expect them to do this in the face of dynamic change in the region's economy.

Another problem with this approach is that it is much too aggregative. The basic–non-basic split, rather arbitrary to begin with, does not capture interrelationships between sectors, or changes in them over time.

This is no problem for the regional I–O models, which are explicitly concerned with the disaggregated structure of production: how much of each of a variety of separate inputs are required for an increment to some regional output? Given a knowledge of these technical production relations, it is possible to determine output in each sector consistent with a new bill of final demands and supply of the region's "primary input", labor. There are, however, a number of problems with the regional I–O approach. To begin with, final demand, though disaggregated, is determined exogenously. Clearly, we would prefer that demands for goods and services in the region be determined endogenously, in response to the proposed new development and the changes in the economy it triggers.

Another drawback of these models is that the I–O coefficients, reflecting the amounts that industries in the region buy from other industries in the region, are fixed. National interindustry models have been criticized for this reason, but the problem is even more serious on a regional level, since movement of firms and industries into or out of the region will almost certainly affect the (assumed fixed) coefficients.* This is noted also by Howe and Easter (p. 58). And the search by firms and owners of resource inputs (including labor) for higher returns in turn ensures that this movement will be a pervasive feature of the region's

* Both problems – exogenous demand and fixed coefficients – also beset interregional input–output models. In addition, interregional models are hampered by a lack of interregional trade data.

economic landscape. Ideally, this sort of maximizing behavior ought to be explicitly modeled.

A final – and perhaps most serious – disadvantage of the I–O method is that it sheds no light on the dynamic adjustment of the economy to the new equilibrium level and composition of output. But this process of adjustment may itself be crucial in studying the effects of a major construction project like an IWT. Perhaps the heaviest impact, for example, on a region's economy and public finances (taxes and expenditures) will come with the early construction phases, and not with the later operation of the project.

These observations have been implicitly directed to the impact on the region benefiting from the water transfer. They obviously apply as well to the other relevant regions, those losing water, those through which water is transported, and those competitive with the region gaining water. But the econometric model I am going to propose as an improvement over the foregoing methods, including regional I–O, is an improvement in fact in part because it can do a better job of assessing the impacts on these other regions. As Howe and Easter put it, "the use of state input–output models precludes any industry-by-industry analysis of impacts outside the states directly affected by the transfer project, so impacts external to the region must be analyzed in *ad hoc* ways" (p. 58). What is wanted, then, is a method for assessing simultaneously, and with equal rigor, the changes in all affected regions. And as noted earlier, it ought to be able to both trace the dynamics of these changes, and relate them to maximizing behavior by private economic agents (or a planner).

A regional econometric model

The essential features of a method, or model, that holds the promise of satisfying these conditions, can be set out briefly as follows. First, it should be *recursive*. That is, forecasts for period t should be made on the basis of data for the previous period, $t - 1$. Then the t forecasts become the input for forecasts for $t + 1$, and so on. This allows us to trace the time paths of the economic activities in a region, including their adjustments to developments like the construction and operation of an IWT.

Second, the model ought to be disaggregated by (economic) sector and region. That is, we are interested, as in I–O, in the behavior of each of a number of key sectors in a region's economy: energy production, other manufacturing, transportation, agriculture, and so on. But – and this is important – we are interested in the behavior of each of these sectors, and the employment in them, in *all* affected regions, not just the one gaining the water.

Third, and perhaps most important, the model ought to be driven by some sort of maximizing behavior, whether we ascribe it to private economic agents or a social planner. That is, the changes in output and employment by sector and region from period to period ought to reflect some attempt to maximize returns.

Putting it all together, we can write a set of forecasting equations like

$$\Delta Q_{ij}^{t} = f_i(\mathrm{TC}_{Xij}^{t\text{-}1}, \mathrm{TC}_{Mkj}^{t\text{-}1}, W_{ij}^{t\text{-}1}, R_j^{t\text{-}1}, K_{ij}^{t\text{-}1}) \qquad (3)$$

$$i = 1, \ldots, n$$
$$j = 1, \ldots, m$$
$$k = 1, \ldots, l$$

where ΔQ_{ij}^{t} represents the change in output, in *value* terms, from period $t - 1$ to period t in sector i in region j; $\mathrm{TC}_{Xij}^{t\text{-}1}$ the transport cost (in $t - 1$) of shipping a unit of output i from region j; $\mathrm{TC}_{Mkj}^{t\text{-}1}$ the cost of obtaining (in region j) a unit of input from sector k; $W_{ij}^{t\text{-}1}$ the wage rate in sector i in region j; R_j^{t-1} the rental price of land in region j; and K_{ij}^{t-1} the existing undepreciated capital stock in sector i in region j. Equations (3) obviously represent a highly simplified version of a multi-region multi-sector forecasting model. There might, for example, be more input prices specified – for different types of labor, for capital if interest rates exhibited any regional variation, and so on. Also, agglomeration variables, such as outputs of major buying and supplying sectors in region j, or measures of congestion, could be significant. But equations (3) do, in my judgment, capture the essential features of regional economic activity and the changes in it. To get a measure of the change in *aggregate* economic activity within a region, we simply take the sum $\sum_i \Delta Q_{ij}^t$, i.e. the sum of the changes over all sectors i.

This change in aggregate activity, or regional product, reflects *all* of the direct and secondary benefits and costs, as defined in equations (1) and (2), to each affected region. And note that regions losing water or competitive with the region gaining are treated on the same basis as the region gaining.

Let us now look more closely at equations (3) and describe the expected relationships between the variables. It is clear, first of all, that output changes ought to be negatively related to all of the input prices, including transport costs. A decrease in any one of these prices, all others held constant, ought to lead to an increase in the change in output. The other (non-price) independent variable in the model, the existing capital stock, is included to reflect the importance of depreciation of existing plant and equipment to a decision on location of production. It ought to be *positively* related to the change in output; given input prices, the larger the fixed investment, the larger the expected increase in output at a particular location. Conversely, the smaller the fixed investment, the more "footloose", or responsive to changes in regional input prices, a firm or industry can be.

Just as the set of equations in (3) represents the changes in output by sector and region, changes in employment (and therefore population) in a region, also presumably of interest to planners and policy-makers, can be represented by a set of equations like

$$\Delta L_j^t = f_j(W_j^{t-1}/W^{t-1}),\ j = 1, \ldots, m \tag{4}$$

where ΔL_j^{t-1} is the net migration of labor into region j from period $t-1$ to period t, W_j^{t-1} is the average wage in region j, and W^{t-1} is the average wage in the (national) economy. We would expect the relationship between the wage ratio and net migration to be positive, to reflect the search by individuals for better earnings opportunities. This is the basic relationship that has been used to explain such familiar patterns of migration as those from Europe to the US in the 19th century, from the south to the north and west in the US for most of the past century, from southern Europe to northern Europe over the past couple of decades, and so on.

Again, an equation like one of those in the set (4) is probably too simple for actual estimation. Moreover, I have said nothing about the form of the functions f_i and f_j, about how regions and sectors ought to be disaggregated, or – most important – how the required data are to be obtained. But at least the elements have been set out of a model which (a) disaggregates by sector and region, (b) gives equal attention to all affected regions, (c) allows for changes in the structure of each region's economy, (d) relates the changes to economizing behavior, and (e) traces the path of the changes over time.

Further elaboration of such a model, and the prospects for implementing it in a study of the economic impact of an actual IWT are beyond the scope of this paper. But those interested in the subject of regional econometric forecasting can consult the work of Harris,[3] in particular. He has in fact developed a model for the US – and more recently for Canada – along the lines just hinted at in equations (3) and (4), but in much richer detail. The Harris model has not been used to assess the impact of an IWT, but clearly it, or something similar, could be, along the lines of the applications to a variety of other resource development projects (see Krutilla and Fisher[4]).

Before proceeding to consider some aspects of the thus far neglected environmental costs of an IWT, let me very briefly indicate how a regional econometric forecasting model like that suggested in equations (3) and (4), or developed by Harris, might be used to measure the conventional economic costs and benefits. The idea is to specify, exogenously, the "primary" activities, such as the construction and operation of a water transfer facility. These activities are then fed into the model, resulting in changes in regional input prices, which in turn trigger output shifts. For example, a lower price of water will lead to an expansion of water-intensive activities within a region. And the expansion of these activities can enlarge the market for still other activities – recall the suggested agglomeration variables like output of major buyers in a region – triggering still further output shifts. Of course, not all changes occur overnight. There is a construction schedule for the project, and the outlays on it, and only some fraction of an industry will move into or out of a region in any one period in response to these outlays and their effects – recall the influence of fixed investment. But presumably the search for higher returns motivates some movement – some change in output – in each period. Once again, the process is much more complicated than I have been able to indicate in these brief remarks, and the interested reader is urged to consult the seminal work of Harris, or some of the applications.

3. ENVIRONMENTAL COSTS OF INTERBASIN WATER TRANSFERS

The basic Howe–Easter benefit–cost relationships, inequalities (1) and (2), provide a framework for consideration of environmental effects of IWT's. There is nothing in these expressions about environmental effects, and indeed, nothing in the ensuing calculations of the direct and secondary benefits and costs of some specific transfers – though the possibility of such effects is noted (pp.106–107). And environmental effects are not treated in the two other excellent comprehensive studies of the economics of IWT's of which I am aware, those by Hartman and Seastone[5] and Cummings.[6] Yet there is no reason why these effects, increasingly recognized as potentially serious, cannot be included in one or another of the cost terms.

One obvious possibility would be "secondary costs", which Howe and Easter indicate arise "through the existence of failures of the market mechanism" (p. 27). Since environmental side effects of various resource development projects are among the outstanding examples of market failure in recent years, the associated costs could be considered a component of secondary costs. But Howe and Easter also define secondary costs in terms of foregone (money) "incomes of factors of production" (p. 26). For example, if the resources employed in an activity displaced by the IWT are not mobile, i.e. cannot move quickly to an alternative, their loss in income is a secondary cost. This definition then looks only at conventional economic costs – even though they are attributed to market failures.

It might be desirable, then, to break out environmental costs separately. This we can easily do by adding a term, "EC", for external, environmental costs, to the right hand side of inequality (1). It would also be added to the left hand side of inequality (2), as an "EC_A", for environmental costs of the alternative (to the water transfer), would be added to the right hand side. If one were interested in the distribution of these costs among the affected regions, they could be entered separately, as EC_X, EC_M, and so on as appropriate.

In principle, then, there is no problem in accounting for the environmental effects of an IWT in the economic calculus. But in practice there is of course a problem; or rather two problems. First, the physical effects must be determined. Second, perhaps more difficult, an economic valuation must be put on them (if, that is, we wish to account for all of the project's effects in a common metric).

About the physical effects I don't claim to know very much. Other participants in this conference, expert in these matters, will be addressing them. But I gather that they can be both substantial and difficult to determine and evaluate. Apparently, changing the water regime of a region can have an effect on its climate, due to greater or lesser evaporation, formation of cloud cover, and so on. And in addition to these micro-climatic effects, certain diversions of water, in particular the very large diversions from Arctic regions to the south now being contemplated in both North America and the USSR, can have an effect on global climate and environment. In one plausible scenario, a reduction of fresh water flow into the Arctic Ocean could lead to a melting of the Polar ice cap, with profound consequences for low-lying coastal areas around the world (Inadvertent Climate Modification: Report of the Study of Man's Impact on Climate, pp. 159–162).[7]

Needless to say, economic evaluation of such effects would not be easy. In some cases, where there are determinate effects on particular economic activities, such as agriculture, evaluation would be feasible. But where it is not, a useful strategy for assessing an IWT might rely on the notion of *dominance* (Fisher and Peterson[8]). This has been helpful in assessing at least one other development project with important, but hard to monetize, effects on the environment: the Trans-Alaska Pipeline (Cicchetti[9]).

Briefly, the notion of dominance is as follows. Suppose two projects, an IWT and one other, say pumping of a groundwater reservoir, can yield the same water output. Suppose further that the costs of each can be broken into two parts: conventional economic costs, measured in money outlays on the required inputs, and environmental effects, measured in various physical units. Let C_m^I and C_e^I represent the conventional and environmental costs of the IWT, and C_m^G and C_e^G the conventional and environmental costs of the groundwater alternative (C_e^I and C_e^G can of course be vectors containing several elements). Then if $C_m^G <$ C_m^I and $C_e^G < C_e^I$, we say that the groundwater alternative dominates the IWT. To compare them it may not be necessary to aggregate conventional and environmental costs in the same metric.

3. UNCERTAINTY AND IRREVERSIBILITY

One final set of issues I wish to address here has to do with the problems and implications for benefit–cost analysis of (environmental) effects of IWT's that are sufficiently long-lived as to be considered irreversible, yet (as with all such effects) not perfectly predictable. What I shall show is that the presence of such effects leads to some presumption in favor of refraining from the activity that gives rise to them. Recall that the basic economic efficiency criterion for an IWT, as given in expression (1) of Section 2, is that the benefits exceed the costs (all properly discounted), or that the *net* benefits be positive. Below I derive a more "conservative" efficiency condition, namely that the net benefits must exceed some positive number.

Since this is a fairly strong result, we ought to be clear about the assumptions which underlie it. First, it is assumed that the environmental effects are uncertain. This seems a very weak assumption; indeed, the converse would be hard to motivate. Second, it will be assumed that the passage of time reduces the uncertainty, in a sense to be defined precisely below. This seems plausible enough, though perhaps not in the rather strong form in which it will be made. Third, perhaps most important, and at the same time most questionable, is the assumption that the effects are irreversible. This may be plausible for certain types of water transfers. One that comes to mind, already mentioned in the preceding section, is the substantial diversion of fresh water flow into the Arctic Ocean, resulting in a reduction in the Polar ice cap, resulting in turn in inundation of low-lying coastal areas, including many of the world's cities. I don't know how likely this is, or whether there are other, more localized effects of IWT's that can be considered irreversible. Other participants in this conference will be addressing these questions. But let me proceed, on the assumption that such effects *are* possible, or even likely, to trace out some implications for policy.

*A sequential model of irreversible investment in an uncertain environment**

Let W_1 be the fraction of a large IWT developed in the first period and W_2 be the fraction developed in the second (and last). Let b_1 be the benefit, *net* of environmental costs, from developing the entire project in the first period and b_2 be the benefit from developing in the second. Assume b_1 is known at the start of the first period and b_2 is a random variable with known distribution $b_2 = \alpha < 0$ with probability p, $b_2 = \beta > 0$ with probability $q = (1 - p)$, and expected value $E(b_2) > 0$. The decision problem is how to choose W_1 to maximize the expected value of the project if it is known that the development is irreversible.**

Assume, first, that no further information about b_2 will become available before the start of the second period, when W_2 must be chosen. Since $E(b_2) > 0$, $W_2 = (1 - W_1)$ in any case. The decision rule for W_1 is: $W_1 = 0$ if $b_1 < 0$, $W_1 = 1$ if $b_1 > 0$. This is of course perfectly consistent with inequality (1).

Now assume that b_2 will be known at the start of the second period. If $b_2 = \alpha$, $W_2 = 0$, and net benefit in the second period is αW_1. If $b_2 = \beta$, $W_2 = 1 - W_1$ and the benefit is β. The expected value, at the start of the first period, of benefits over both periods, is then $b_1 W_1 + p\alpha W_1 + q\beta = W_1(b_1 + p\alpha) + q\beta$. This expression is to be maximized by an appropriate choice of W_1. Since the expected value criterion is linear in W_1, the decision rule is again of the "bang-bang" type: $W_1 = 0$ if $(b_1 + p\alpha) < 0$, $W_1 = 1$ if $(b_1 + p\alpha) > 0$. But note that, since $p\alpha < 0$, this rule is clearly more conservative than the previous one. Now b_1 is required to exceed some positive number, $p\alpha$, whereas previously it was required only to exceed zero.

The point of this exercise has been to show that the accumulation of new information (reduction in uncertainty) about a project that will have an irreversible impact on the environment implies that the project's expected value will be maximized by a relatively conservative decision rule, one that puts a greater "burden of proof" on the project. Or in other words, there is some presumption in favor of refraining from it. Note, however, that just because something is irreversible it should not, on that account, not be undertaken. The rule just derived is more flexible. It says that the combination of irreversibility and (reduction in)

* A more general statement of the model can be found in Arrow and Fisher[10].

** Maximization of expected value is perhaps the simplest decision criterion. Others, more complicated, involving one or another variant of risk aversion, would not change the results obtained below. On the contrary, the results would be obtained more easily.

uncertainty in effect gives rise to an additional, but finite, cost of a project. The economic efficiency condition is a modified form of inequality (1). The project's net benefits, i.e. the left hand side of (1) *minus* the right hand side, must exceed this new, positive, "cost" term, rather than zero.

5. SUMMARY AND CONCLUSIONS

With recent increases in the size of proposed IWT's, careful consideration of their economics becomes particularly important. In this paper I have put forward a number of proposals and propositions concerning the theory and measurement of the costs and benefits of an IWT. They can be restated briefly as follows.

(a) Commonly used methods of measuring the conventional economic impact, including input–output analysis, are not entirely adequate, in that they do not allow sufficiently for induced changes in the structure of the economies of the impacted regions, do not trace these changes through time, and do not relate them to maximizing behavior by economic agents. In Section 2 I propose an econometric modeling approach that might accomplish these objectives.

(b) Calculations of the benefits and costs of an IWT ordinarily ignore its effects on the environment, yet these are likely to be substantial. In Section 3 I indicate how the standard decision criterion should be modified to include the costs of environmental effects. Where the costs cannot be estimated, I suggest a technique for comparing an IWT to an alternative means of producing water, that still accounts for both conventional economic and environmental effects of each.

(c) It is possible that the environmental effects of an IWT may be both irreversible and uncertain. Where, however, the uncertainty diminishes over time, as better information about the effects and their costs becomes available, I show that there is a kind of additional cost to proceeding "too soon" with the project. This represents a further modification of the standard benefit–cost criterion.

REFERENCES

1. Howe, C.W. and Easter, K.W. *Interbasin Transfers of Water: Economic Issues and Impacts,* Johns Hopkins Press, Baltimore, MD, (1971).
2. Isard, W. *An Introduction to Regional Science,* Prentice–Hall, Englewood Cliffs, NJ, (1975).
3. Harris, C.C. *The Urban Economics, 1985: A Multiregional Multi-Industry Forecasting Model,* Lexington Books, Lexington, MA, (1973).
4. Krutilla, J.V. and Fisher, A.C. *The Regional Economic and Fiscal Impacts of Large Scale Energy Resource Development: Northern Great Plains Coal,* manuscript, Resources for the Future, (1977).
5. Hartman, L.M. and Seastone, D.A. *Water Transfers: Economic Efficiency and Alternative Institutions,* Johns Hopkins Press, Baltimore, MD, (1970).
6. Cummings, R. *Interbasin Water Transfers: A Case Study in Mexico,* Johns Hopkins Press, Baltimore, MD, (1974).
7. *Inadvertent Climate Modification: Report of the Study of Man's Impact on Climate,* M.I.T. Press, Cambridge, MA, (1971).
8. Fisher, A.C. and Peterson, F.M. The environment in economics: a survey, *J. Econ. Lit.,* **XIV** (1), (March 1976).
9. Cicchetti, C.J. *Alaskan Oil: Alternative Routes and Markets,* Johns Hopkins Press, Baltimore, MD, (1972).
10. Arrow, K.J. and Fisher, A.C. Environmental preservation, uncertainty, and irreversibility, *Q. J. Econ.* **88** (2), 312–319 (May 1974).

Water Supply & Management, Vol. 2, pp.147-157. 0364-7714/78/0601-0147$02.00/0

The Water Grid Concept

ROBIN R. REYNOLDS*

ABSTRACT

Phases in the history of the development and use of water resources are reviewed. A possible ultimate pattern of such development for a nation or a large international region is discussed. It is suggested that some insight into the characteristics of one such possibility of an ultimate phase can be gained by considering the characteristics of a large power grid system. On this basis, the characteristics of a water grid are described. The most significant characteristics are large interbasin and interregional aqueducts and a central coordination and management.

I. INTRODUCTION

The foundation of our civilization is the human development of two primary natural resources: land and water. Our early views of man are of him emerging from the shadows of the Stone Age as a cultivator of soil and an applier of water. The story of the development of that water from the earliest times to the present has many chapters.

Phases in the history of water development

In the early phases of development of regions, the demands on the water resources generally, were negligible compared with the quantities available. As regions became settled and permanent communities were established, the region's water resources were put to use to meet human and livestock needs, to meet the needs of communities and settlements and, in the arid regions of the world, to irrigate agricultural lands. The first irrigation developments were shallow wells and diversions from flowing streams for immediately adjacent uses. This was the first pattern of water development; and in many parts of the world, it is still the only pattern of water development.

Development of surface-water supplies – As diversion of water from a stream increased, a point was reached when the natural flow of the stream was at times insufficient to meet the needs. Reservoirs were then constructed. The first such reservoirs were generally single purpose, supplying water to a farming area or to a community. With further increases in demands on the water supply, larger reservoirs were needed to increase the quantity of water that could be made available by providing cyclic storage where run-off was carried over from wet portions of the year into the dry portions, from wet years to dry years, and from wet cycles to dry cycles. The larger reservoirs also served multiple purposes. They provided storage space which could be held empty in reserve to impound flood flows and they could be operated to serve other purposes such as to produce hydroelectric power and to maintain more uniform

* Deputy director, California Department of Water Resources, Sacramento.

flows in downstream reaches for navigation, fish preservation and enhancement, and to provide water quality control. Not only were water supplies put to greater use, but the economic and financial bases of projects were broadened by providing for such multiple uses.

Development of groundwater supplies – As the settlement of regions using groundwater proceeded and the water needs increased, the number of wells was increased as well as their depth. Drilling of deeper wells was made possible by greater technical knowledge and by improved drilling equipment. Significant advances also have been made in well construction.

An appropriate number of deeper and improved wells combined with management of pumping and recharge made possible a greater utilization of the groundwater reservoir. In theory, a groundwater reservoir can be operated in much the same manner as a surface reservoir with both annual and long-period cycles of filling and drawdown.

Coordination of use of surface- and groundwater supplies – In most of the irrigated areas of the world, the development and use of both surface-water and groundwater recources is the rule rather than the exception. The development and use of each is, however, generally independent. But a next and logical step is their coordination, and in some groundwater basins this is being done. Full coordination of the operation of surface- and groundwater supplies requires special management and legal arrangements. The objective of such an operation is to make the most economic use of a basin's surface- and groundwater reservoir volumes to regulate the total basin water supply for all uses.

Water conveyance facilities – In addition to surface- and groundwater storage reservoirs to conserve and regulate the water supplies, it is necessary to construct canals and aqueducts to convey the regulated water to the area or place of use. In the first stage of development, a canal was constructed to bring water from a stream to a farm, then from a river or reservoir to a larger area of use, and finally long aqueducts were constructed to carry water from a remote source of supply. The large aqueducts are often interbasin and a few are interregion. They transfer water from areas with excess water supplies to those deficient in water supplies.

Management of the use of water – Man has learned to regulate natural water supplies as they occur with respect to time, in order to have them available at the time they are needed for human use. He can build conveyance systems to move the regulated water supplies to the places of use. More recently he has learned how to manage the removal of the accumulating saline drainage waters which brought ruin to early irrigation projects and which are still a problem to all. He is also learning about water quality control. No water is good for every use, but every water is good for some use and the quality of the water should fit its use. As it is an objective to have the most economic balance between use of surface- and groundwater reservoirs, so it should be an objective to make the most economic use of water supplies from a water quality standpoint. Higher quality should be used where it is required and lower quality water used where it can be tolerated. Improvement of water quality by treatment should also be considered.

II. WHAT IS THE ULTIMATE PHASE OF WATER DEVELOPMENT?

Section I discussed phases in the history of the development and use of water resources. All of those phases exist today in the world. In fact, in most nations water development is in its early phases. In only a few areas is there beginning to be a coordination of the operation of surface- and groundwater reservoirs. Water quality control is beginning to be considered, but only where pollution makes it absolutely necessary. There are very few interbasin or interregion aqueducts to balance water supplies between areas of surplus and areas of deficiency on a large scale.

There is much remaining to be done to develop the world's water resources for all uses. It is suggested that it would be worthwhile to consider the possible phases that might exist in the future or to consider what might be an ultimate phase of water development. A general description of such an ultimate phase would be a useful guide to water resources planners and managers. Its primary value would be as a guide so that interim steps taken today would fit into the next phase and into the ultimate phase.

Objectives of the ultimate phase of water development – As a first step in attempting to describe such an ultimate phase of water development, the objectives should be set forth.

What should be the objectives of the ultimate phase of development of the water resources of a nation or a large international region?*

Such an ultimate development should:

(1) meet all human needs including those of nomadic people and their livestock;

(2) meet water needs for all purposes:

- (a) villages and communities,
- (b) agriculture and livestock,
- (c) city and municipal,
- (d) industrial,
- (e) power plant cooling,
- (f) hydroelectric power,
- (g) navigation,
- (h) fish and wildlife,
- (i) recreation,
- (j) environment,
- (k) esthetics.

(3) The foregoing needs should be met as they occur and particularly in dry portions of the year, in dry years, and in dry periods of climatological cycles.

(4) The foregoing needs should be met by water of adequate quality, both mineral and biological.

(5) Human developments and natural resources should be protected against damage by floods.

(6) To meet the foregoing needs, water supplies available in all phases of the hydrologic cycle should be considered.

(7) Adequate drainage should be provided.

(8) Constructed facilities and their operation and management should be such that all of the foregoing needs and requirements are met in the most economic manner.

Development of other natural resources

The world has many other resources, in addition to water, which man has developed. For example, the human, animal, mineral, vegetative, power and ocean resources all have been put

* In the discussion in this paper the term "region" is often used. It is assumed to be a large area that might transcend national boundaries.

to use by man to meet his needs. In looking for guides to future phases of water development, the patterns of development of these other resources should be considered. One of these, power, has some attributes similar to water. It is made available at specific points, it requires distribution, needs should be met on demand, and for efficiency and economy its facilities require coordination. It is these similarities that suggest a concept of water development similar to a power grid system where the power needs of a region are met on demand by integration and coordination of resources and facilities and where large-capacity power transmission links are the backbone features of the physical facilities.

III. CHARACTERISTICS OF A POWER GRID

The term "Power Grid" is a familiar one. Power grids, often international, exist today in the developed nations and are an objective of the underdeveloped nations. Today the emphasis in electric power systems is on coordination between systems. This trend began in the late 1920s when strong interties began to be built to provide economy and improved reliability of service. With the interties, separate electric systems could join into increasingly larger power pools.

Many benefits resulted, including larger generating units, savings in providing reserve capacity, exchanges to take advantage of diversity between systems, and the ability to select the most favorable generating sites.

The development of electric power interties has resulted in the construction of lines to transmit large blocks of power in the extra high voltage range. A 345 kV line was constructed in the United States in the early 1950s. In Sweden a 400 kV line was placed in operation in 1952, and in the USSR in 1954. The voltage on the Soviet lines was increased to 500 kV and then to 750 kV. Such interties are the backbone of the world's power grid systems. They are super highways of low cost power.

A power grid is an interconnected power system which serves, on demand, the needs for electric power including, for example, domestic, municipal, industrial, transportation, manufacturing and agricultural needs. In a power grid, electric power from various sources can flow on alternative routes to the many points of use either under a planned operation or in unforeseen circumstances of outages of production facilities, of transmission systems and of demand loads, and there usually is an excess capacity to meet unexpected needs.

IV. CHARACTERISTICS OF A WATER GRID

If a power grid is used as an analogy to define a water grid, a water grid could have the following characteristics:

(1) It could be a conveyance system that conveys water from water sources to places of water use.

(2) It could have alternative sources and alternative routes in the event of the primary sources or routes being out of operation.

(3) It could have interconnecting links, so that diversity in the availability of water among different water sources and in demand for water among different service areas could be taken advantage of to make maximum use of the available water resources and to obtain the most economic use of facilities.

(4) It could have sufficient conveyance capacity to meet peak demands as they occur

with some excess capacity.

(5) It could be connected to, and convey water from all sources including those sources under the various phases of the hydrologic cycle and those sources where water is made available by technological processes such as desalination.

(6) The operation of the facilities of the grid system could be coordinated and there could be an integrated management.

In summary, the water grid could link the water sources to the areas of water demand. It could provide a physical system to convey water to meet needs under the various conditions of demand and of availability. It would transmit various quantities and qualities of water from available natural sources in the hydrologic cycle together with supplies from technological developments such as desalination, waste water reclamation and weather modification. Finally, it could be a complex and frequently large scale interbasin or interregion water transfer system transcending physical boundaries with an integrated operation of facilities and a centralized management. It is these two characteristics, large water transfer facilities and an integrated operation of facilities and centralized management, that would be the primary characteristics of a water grid.

Specific objectives that a water grid would need to meet

There are specific objectives and operational requirements that a water grid would need to meet. A number of these are set forth in the following sections. They have been developed using the power grid as an analogy by which to visualize a water grid and in addition by introducing some concepts of a power grid into our water management thinking and by introducing some recent and new concepts of water management.

Variations in water demand – Depending upon a number of factors, including the weather, type of use, rate of development, etc., demands for water vary during the day, during the week, by months, and from year to year. A water grid system operating in conjunction with water conservation facilities should have adequate capacity and sufficient operational flexibility to meet water needs under such variations. The daily and weekly variations are usually met by adjusting releases from reservoirs and by withdrawing water from or adding water to the aqueduct facilities. On the other hand, longer term variations in demand occur during a year. For example, agricultural demands are high during the growing season and low during the nongrowing season while municipal and industrial demands are more constant through the year. These variations are met by adjusting releases from storage reservoirs.

In sizing of facilities to meet such variations in demand, it is important to consider the peak demands that will occur under conditions of full development. As can be seen, two factors are important. The conservation and transportation facilities must have adequate capacity and there must be operation flexibility.

Water quality management – In the same way that adequate quantities of water must be made available, water of adequate quality, both mineral and biological, must be made available. In a water grid system water quality monitoring and management would need to be carried out so that water of various qualities would be utilized appropriately throughout the system depending upon the types of need and the physical characteristics of the service areas.

Alternative routes – At the present time there are very few examples of alternative routes in major water conveyance systems. This is probably because canals and aqueducts operate reliably and because of the large cost involved. When an aqueduct is constructed linking a water source with an area of demand, it is difficult to justify construction at the same time, of a second aqueduct along an alternative route. As water demands increase, however, and it becomes necessary to construct a second aqueduct either to the original service area or to a new service area, and an aqueduct extending in a general direction parallel to the first is envisioned, it would be important to consider the advantages that might result if the second aqueduct were built on another alignment and if some additional capacity were constructed in the second aqueduct so that one could be out-of-service either from emergencies or as a planned operation, for example, to relieve peaking needs. An interconnection or interconnections between the two would also be needed.

Interconnections – As just mentioned, to get maximum use from generally parallel aqueducts along alternative routes, it would be necessary to have interconnections. Interconnections also would increase the operational flexibility of existing systems, and such systems should be reviewed looking for opportunities to make interconnections.

Consideration of reversible flow – An important characteristic of a power grid is that power can flow in either direction. Such a concept is not so applicable to water conveyance systems but the possibility that it might prove useful should not be overlooked.

There are a number of possible situations where an aqueduct with this capability would be useful. Such a situation would exist when weather conditions resulted in there being excess water supplies at some point or points along an aqueduct and there was a need for water at upstream locations. An aqueduct that could carry water in either direction would also be useful as a connection between two or among several adjacent basins in order to move water that was excess to needs in one basin, to basins where water supplies were deficient.

It would be possible to design, construct and operate an aqueduct to provide for reversible flow. Large aqueducts are constructed with very small slopes. For example, a concrete lined aqueduct with a capacity of about 300 m^3/sec (about 11,000 ft^3/sec) has a slope of about 1 in 25,000. Such an aqueduct can convey large quantities of water with only a small loss of head. This gives rise to several possibilities if reversible flows are desired. Aqueducts of this size could be constructed at no grade, i.e. level, and the hydraulic head could be created by pumping stations at both ends of such an aqueduct. Intermediate pumping lifts with reversible pumps also could be utilized.

To provide reversed flows in existing aqueducts, either temporary or permanent pumping facilities could be constructed. For each of these possibilities in a large aqueduct of the size stated, a pumping lift arrangement providing a lift of 1 m (3.3 ft) for every 25 km (15 miles) would be necessary.

The same possibilities exist with smaller aqueducts but the distance between pumping stations would be less or higher lifts would need to be provided.

Groundwater basins as parts of the water grid

The most important function of a groundwater basin is to store water. It has the advantages of not requiring any land area and it has no evaporation. In addition, a groundwater basin has several physical characteristics which are analogous to characteristics of a power grid. It is important to recognize these in considering the functioning of groundwater basins as parts of a water grid system. A groundwater basin provides alternative routes from areas of supply to points of use, it provides interconnections and it will allow reversible flow. These characteristics can be used in the management of both water quantity and water quality.

Possible arrangements to give operational flexibility

So far the discussion has generally related to physical facilities. An important concept of the water grid is the coordinated management of the physical facilities. The objective of such coordination would be to give complete flexibility of operation so that the greatest overall economic benefit for the least economic cost would result.

To attain the most economic operation, water needs would be met by releasing appropriate amounts of water from the most appropriate reservoirs. Determination of the amount to be released from each reservoir should not be based upon ownership or rights to the water but should consider the needs to be met, the capabilities of the facilities, the amounts of water available and the costs and values involved. In a water grid system with integrated management, all water in the system and all facilities of the system would be used to the maximum and in the most economic manner. Reservoir spills should be avoided and all possible discharges should be through power plants. Releases for managed river flows should be from those reservoirs where abundant water supplies are occurring. Diversions to aqueducts also should be from streams and reservoirs where there are abundant water supplies. Conveyance of water to places of use or to terminal reservoirs for later use should be by the most economic route. All conveyance facilities should be used during wet periods to avoid loss of water.

Following are some suggestions for possible exchanges and possible operation and management plans to make maximum use of water supplies and to get maximum operational flexibility and performance from the total physical system. When exchanges and banking of water and power and exchanges of rights to the use of physical facilities are discussed, it is pointed out or is to be understood that it is necessary to keep accounts of such exchanges and banking and often of the values being exchanged or banked. This is necessary when the grid system consists of a number of smaller systems being coordinated as a single larger system so that the rights and values of these smaller systems are preserved. It also is necessary for a single large system in order to attain the most economic operation.

Exchange of water in reservoirs – Contractual or management arrangements should be worked out so that exchanges of water, in an ownership sense, between and among reservoirs are possible. This will require a system for accounting for the amounts of water exchanged and of the value of the water on some common base for the grid system.

Exchange of capacity in reservoirs – Rights to the use of the capacity in the grid system reservoirs should also be able to be exchanged. This will also require an accounting of the capacity so exchanged and its value.

Exchange of water in aqueducts – Exchange of water flowing or stored in an aqueduct should be provided for. This will require an accounting of the water and its value.

Exchange of capacity in aqueducts – Exchange of rights to use the physical capacity of aqueducts should be provided. This will require an accounting of the capacity so exchanged and its value.

Power exchange – In addition to the reservoirs and water conveyance aqueducts, the facilities of the water grid system will usually include hydroelectric power plants and pumping plants. The electric capacity and the electric energy production capability of the power plants are important system resources. The pumping capacities of the pumping plants are also important system resources and the energy required for pumping plant operation is a significant system requirement. Exchanges in the use of power plant and pumping plant capacities should be provided for. Exchanges of the energy produced by the power plants should be provided so that the power needed for system pumping or to meet other system obligations is furnished from the plant of the system that is the most economic for the particular situation.

Power and water banking – In the coordination and integration of power and water systems, the terms "exchange" and "banking" have special meanings. The use of these terms is not always consistent among systems. For the purpose of this discussion, the term "exchange" when used in the accounting for water, covers not only operational exchanges that might be made on an hour-by-hour or day-by-day basis, but also exchanges over a long time period, possibly as long as a year. When electric power is exchanged on an hour-by-hour or day-by-day basis, the term "exchange" is used. When, however, the exchange of electric power is for a period longer than one day, the term "banking" is used. In operation of the water grid system exchanges of water and power should be accounted for and banking accounts should be provided for accounting of long-term exchange of electric power among facilities or among segments of the system.

Exchange of use of facilities – In the operation and management of a water grid system, the concept of the exchange of use of facilities is an important concept. In previous sections of this paper, the need has been discussed to provide for the exchange of the use of capacity in reservoirs, for exchange of the use of capacity in aqueducts, and for the exchange of the use of capacity of power production plants and pumping plants. Arrangements should be made so that such exchanges can be made in as complete and as flexible a manner as possible. Exchanges in the use of the production capacity of power plants will allow project pumping needs and other obligations of either the entire system, or portions of the system, to be met by the power plant or where the most abundant water supplies are available or by the most economic plant, considering not only the plant characteristics but the distance of transmission. Exchange of the use of capacity in aqueducts and of the capacity of the related pumping plants should be provided for so that water can be conveyed by any of the alternative routes available, depending upon circumstances of operational needs, emergency situations or economic considerations.

On-peak and off-peak operation of generation and pumping facilities – The general character of the daily, weekly and yearly demand for electric power throughout most of the world is generally similar. Daily demands are high during the daylight hours and lower during the night. Demands are highest during the weekdays and lower on weekends. The demands throughout the year do not fall into such consistent categories because there are generally differences in demands because of the uses that are met and because of the different characteristics of summer and winter needs. But the general consequence of these variations in demands and their interrelations is that there is a predictable minimum base load demand that must be met continuously throughout the year and from year-to-year. Demands above this base load are the peaking demands and the power production facilities, which do not operate continuously, meet such demands. Although the power production facilities which meet the peak loads do not operate continuously they must be dependably available. The facilities which meet peak loads are often called on-peak facilities and the period of their operation is called the on-peak period. Since they must be available but do not operate continuously, their costs are higher for each unit of production. Therefore, the value of the power produced during these periods or the cost of power needed during such periods is higher.

These considerations must be kept in mind for the most economic operation of a water grid system. To take advantage of these circumstances, hydroelectric power production facilities should be operated as much as possible during the peaking periods and pumping plant facilities should be operated as much as possible during the off-peak periods.

Operation to maximize power production and operation to maximize water yield – A given reservoir with its related power production facilities can be operated under an infinite number of operation plans. These plans fall, however, into two general categories – an operation which will maximize power production, or an operation which will maximize water yield. Although multi-purpose reservoirs are operated to meet many other demands, such as those to provide flood control and to provide minimum flows in the downstream channel, such demands can usually be readily met whether the reservoir is being operated to produce a maximum amount of power or a maximum water yield. Planning of the operation of the grid system reservoirs should be such that each reservoir produces an appropriate dependable power capacity, energy production and water yield so that in the aggregate, all system needs and commitments are met.

Controlled volume concept of aqueduct-operation – Much of the activity and many of the problems in the operation of aqueducts and aqueduct systems relates to adjustments required when changes in flow are made. That activity and those problems are at a minimum when the aqueduct is conveying a constant unchanging quantity of water. When that flow is changed, adjustments of all control facilities are required. The more frequently such changes are made, the more frequently adjustments are required with consequent higher costs. In addition, for aqueducts of considerable length, the time between when the flow change is made at the head of the aqueduct and when it is felt at the lower end is considerable – that time being only somewhat less than the time it takes a particle of water to flow the length of the aqueduct. In other words, the time required to respond to a change in demand in a service area at the end of the aqueduct is long. When agricultural demands are being met, this problem generally is not significant but when municipal and industrial demands are being met, or when emergencies occur, problems can arise.

Modern control system techniques, including the use of computers, have improved this situation substantially. If by the use of such equipment, all facilities along an aqueduct, which generally include pumping plants, check gates, and major delivery turnouts can be operated so that their operation is simultaneous, a much higher degree of control of the operation of the aqueduct can be obtained. For example, an aqueduct with such a remote control system can be brought from a condition of no flow to a condition of full flow in a short period by simultaneously starting all pumping plants and simultaneously opening all check gates. In the same manner, the flow can be brought to a halt by simultaneously turning off all pumping units and closing all check gates. Such a method of operation should be considered for appropriate aqueducts and interconnection links of a grid system to improve service to the users and to allow faster response to emergency conditions.

Long-range forecasts – Long-range forecasts of the operation of grid system facilities should be made for two primary purposes. First, to assure that system facilities are in as good a position as possible to meet demands as they occur under the many possible conditions of demand and water availability that may arise, and second, to assure that additions to and physical changes in the system facilities are made on a timely basis, considering the long lead time for design and construction. Detailed operation studies made on a monthly basis and projections of water needs and operation requirements at least 20 years into the future, should be updated annually. The 20-year or longer projection, and possibly an intermediate projection, should be studied in relation to a long-term water supply period which includes not only critical water supply periods but also is representative of the long-term water supply.

Projections for about the next 5 years, and particularly next year's operations, should be made in great detail, considering at least three possible conditions of water supply – the normal water supply and two extremes, for example, a upper and a lower quartile water supply. If a critical water supply year is actually being experienced, an operation under the possibility that the next year also will be another critical year should be considered.

Need to make computer studies – Studies and implementation of the foregoing concepts of exchanges, banking, alternatives of operation, and operation management require not only a large number of computations but computations in considerable volume. In fact, operation studies of the coordinated operation of the reservoir and aqueduct facilities of a large regional area would be virtually impossible without electronic computers.

System operation models are in existence which allow the study of the coordinated operation of a large number of reservoirs and related aqueducts. For example, the United States Bureau of Reclamation and the California Department of Water Resources have jointly developed a computer model for the entire Central Valley of California. In this model the Central Valley is divided into 40 hydrographic areas, including both the mountain watersheds and the valley service areas. With this model the individual and coordinated operation of the Central Valley's 56 major reservoirs can be studied. Routines for operation of the major power plants also are included.

Projections of future demand for 1980, 1990 and 2020 have been made and each of these future demand projection periods can be studied over a 33-year water supply period, and this is being extended to 51 years.

A separate but related model has also been developed by the California Department of Water Resources not only to study but also to manage the operation of the California Aqueduct which extends nearly 450 miles from the Delta of the Sacramento and San Joaquin Rivers to terminal reservoirs in Southern California. In this study, the aqueduct is divided into 72 separate reaches with the six enroute and terminal reservoirs each being handled separately.

In addition, both analog and digital computer models have been developed to study the flow patterns in some 1100 miles of channels of the Sacramento–San Joaquin Delta with its 50 islands.

All of the foregoing models consider only quantities of water either stored in reservoirs, flowing in the rivers and channels of the Central Valley, or in the aqueducts of the Federal and State projects. A model to study the mineral water quality of the Delta channels and the San Francisco Bay system also has been developed since the lower estuaries and the Bay involve a transition from fresh to ocean water with the problem of saline intrusion modified by tidal flows. One final model, although not yet operational, is being developed to study the biological quality in the Delta and estuary channels. Such models are examples of those that would be needed for a water grid system, depending, of course, on the particular physical situation and the system facilities.

Centralized operation control – In order to optimize the operation of all system facilities and to provide the essential central management, it would be necessary to have a centralized control of the management of the grid system, including its facilities and operation management. Such a central management, however, must be responsive to the total needs and economy of the area being served.

In developing and carrying out the plan of operation, all water demands must be considered, as must be the need to produce hydroelectric power, to provide navigation, to provide water quality management, to protect and enhance the fish and wildlife resources, to provide flood protection for human developments and for natural resources, to provide recreation, to protect and enhance the environment, and to enhance the quality of human life. Input from all of these interests must be provided and must be considered. These interests must be involved in the decision making process.

Water use management

In all the foregoing, the discussion has related to providing for and managing the water supplies to meet water needs. However, the situation should also be studied where water supplies do not meet water needs. In this situation two alternative courses are possible; either additional water can be supplied or demands can be reduced. The objective should be a compromise program which considers both concepts and results in the most economic program.

Water utilization can be improved in a number of ways and this reduces demands with the result that available water supplies can be extended to meet additional uses. Operational losses can be reduced by lining of aqueducts and canals. Irrigation efficiencies can be increased by improved irrigation methods which involve application of smaller amounts of water such as by drip irrigation. Reuse of water is possible either by subsequent use or by reclamation and reuse. Other management methods also are available. For example, the amount of use of water is closely related to its cost. Therefore, revised or new pricing systems could be effectively used to influence the demand for water.

Water Supply & Management, Vol. 2, pp.159-176. 0364-7714/78/0601-0159$02.00/0

Environmental Assessments in Water Resources Planning

LEONARD ORTOLANO*

INTRODUCTION

Since 1970, environmental assessments have been required for water resources planning studies carried out by the federal agencies responsible for water resources development in the United States. The principal reason for this is the National Environmental Policy Act of 1969 (NEPA), a law that requires all federal agencies to describe the environmental impacts of actions they propose to take. As a consequence of NEPA, federal water resources development agencies like the US Army Corps of Engineers and the Soil Conservation Service prepare "environmental impact statements" for proposed water projects, and these statements are reviewed by other agencies, various interest groups and individual citizens.

A more recent requirement for environmental assessments is contained in a set of planning regulations issued by the US Water Resources Council.[1] These regulations, known formally as the "Principles and Standards for Planning Water and Related Land Resources" (referred to herein as the "Principles and Standards"), require an assessment of the environmental effects of alternative actions considered by an agency. The regulations elevate "environmental quality" to the status of a formal planning objective along with the traditional objective of "economic efficiency". In addition, the Principles and Standards require that an alternative action known as the "Environmental Quality Plan" be formulated to demonstrate how water resources planning goals can be met while preserving and enhancing environmental values. Taken together, NEPA and the Principles and Standards have caused the federal water resources agencies to devote a good deal of attention to the ways in which environmental assessments should be carried out.

This paper examines several aspects of the process of conducting environmental assessments. It begins with an overview of the types of environmental impacts associated with water projects. This is followed by a discussion of the collection of procedures that have been brought together under the label of "environmental assessment methods". The portions of the paper that come after the discussion of methods are based on the premise that the principal issues involved in the environmental assessment of water projects do not concern methods *per se*; rather, they concern the ways in which the information generated by the use of these methods is integrated into other activities that are part of the water resources planning process (e.g. the formulation of alternatives). Questions relating to the influence of information resulting from environmental assessments are pursued in two parts. One of these parts concerns research which indicates that the environmental assessments carried out in response to NEPA have not had a great influence on federal water resources decision making. The second

* Associate Professor, Department of Civil Engineering, Stanford University, Stanford, CA, USA.

of these parts, which is also the last major section of the paper, outlines a planning process which fosters the use of information from environmental assessments in various aspects of water resources planning and decision making.

ENVIRONMENTAL IMPACTS OF WATER PROJECTS

Because the term "environmental impact" has acquired several different meanings it is necessary to clarify its usage herein. In so doing, a position similar to the one taken by the US Army Corps of Engineers is adopted. Under the River and Harbor and Flood Control Act of 1970 (Public Law 91–611), the Corps is required to assess the economic, social and environmental effects of any projects it proposes to carry out. As regards economic effects, the Corps has been making such assessments since the late 1930s; these have taken the form of benefit–cost analyses and have been conducted by organizational units typically known as "economics sections" in the various District Offices of the Corps. The environmental and social effects have come to include all impacts that are not considered in a traditional benefit–cost analysis, e.g. air quality degradation and noise pollution. These effects are generally assessed by Corps staff members located in an "environmental section" (or branch). It is this collection of effects (i.e. all impacts except those assessed in a benefit–cost analysis) that receive prominent treatment in Corps' environmental impact statements and that are referred to herein as "environmental impacts".

Examples of environmental impacts: the California State Water Project

To further illustrate the types of effects included under the above definition of "environmental impact", we briefly note aspects of the California State Water Project, a major undertaking involving the interbasin transfer of water. As shown in Fig. 1, the Project carries water from northern California rivers (e.g. the Feather River, the Sacramento River) across the Sacramento–San Joaquin Delta and south via the California Aqueduct to serve the bulk of its users in Southern California.[2] The project is designed to deliver of the order of 4.23 million acre-ft/yr (5.21×10^9 m^3/yr).

The State Water Project has been severely criticized because of the adverse environmental effects it may cause. Gill, Gray, and Seckler have identified three principal lines of criticism in terms of environmental effects.[3] First, it has been argued that the Los Angeles area has "already grown beyond supportable dimensions" and that further growth should not be encouraged by Project water; the premise here is that by providing municipal water supply, an important mechanism for controlling urban population growth is lost. Second, the overall effect of the Project on the Delta and on San Francisco Bay is not known, but it is likely to be significant and adverse. For example, the Project could lead to losses of fishery resources and wildlife habitats and to significant adverse effects on water quality, e.g. excessive growths of undesirable aquatic plants. (A summary of possible effects on the Bay–Delta system is given by Goldman.[4]) The third and final criticism concerns the necessity to supplement the freshwater flows to the Delta as water withdrawals increase there and in the Sacramento Valley; this would likely require impoundment projects on the rivers of the State's north coastal area, namely the Eel, the Klamath and the Trinity. According to Gill, Gray and Seckler, Project critics cite the effects in the north coastal area as including: destruction of "one of the last refuges of nature in California"; destruction of the area's valuable fishery resources; and the accumulation of silt in upstream areas with its attendant effect, the

Fig. 1. The California State Water Project.

degradation of downstream beaches that depend on the rivers' silts to replenish natural beach erosion.

Impacts associated with typical projects

The California State Water Project illustrates some of the effects that are included under the heading of "environmental impacts". A more general overview of the impacts commonly

*Table 1. Types of environmental impacts commonly associated with Interbasin Water Transfers**

Area of impoundment	Along conveyance route
Submerges land area	Using river channels
Modifies aquatic ecosystem (e.g. fisheries, insect populations)	Increases flows and changes groundwater recharge
Modifies terrestrial ecosystem (e.g. wildlife habitat)	Changes water quality and temperature
Changes water quality and temperature	Alters fish production
Increases evaporation and affects microclimate	Changes riparian vegetation
Affects erosion and sedimentation	Modifies erosion and sedimentation
Alters groundwater and geologic features	
Influences land use (e.g. recreation facilities near impoundment)	Using canals
	Interferes with land access
	Destroys fish at intakes
	Decreases wildlife habitat
	Creates safety hazards for children
	Provides opportunities for recreation
Downstream from impoundment	**Area of water use**
Modifies hydrographs	Allows population to grow
Affects groundwater recharge	Accommodates expansion of urban centers with associated effects
Changes aquatic and terrestrial ecosystems	Supports expansion of irrigated agriculture with associated effects
Alters water quality and temperature	
Modifies sediment transport	
Influences land use (e.g. residential development in flood plain)	

* Adapted from Hagan and Roberts[5] and Ortolano[6].

associated with major water projects like interbasin water transfers is given by Hagan and Roberts.[5] They organize their discussion of impacts in terms of geographic location: (1) area of impoundment; (2) area downstream from impoundment or project diversion, or both; (3) area along conveyance route; and (4) area of water use. Table 1 elaborates on this four-part classification by indicating the broad categories of impacts associated with each of the areas.

In addition to the work by Hagan and Roberts, there have been several other general reviews of the environmental impacts commonly associated with water projects. Three such reviews, each focusing exclusively on a particular type of structure or activity, are contained in a report by the Stanford Workshop on the Environmental Impacts of Water Projects[6]: one concerns impoundments, the second concerns channel modifications, and the third concerns dredging and spoil disposal. Reviews of this type can provide a path into the widely scattered literature on environmental impacts; they can also provide the engineers and economists who have been traditionally involved with water resources planning with insights into the broad range of impacts that need to be considered in the course of an environmental assessment.

ENVIRONMENTAL ASSESSMENT METHODS

Having provided a definition for the term "environmental impact" and an indication of the environmental impacts commonly associated with water resources projects, we now consider the methods used in carrying out such assessments. These methods have been the principal subject of several recent textbooks and at least six survey articles or reports.[7]

Following Dickert,[8] we discuss environmental assessment methods in three parts corresponding to impact identification, prediction and evaluation (see Table 2). Methods for identification consist of materials that provide those conducting environmental assessments with general guidance on the types of impacts that *may* be associated with a particular type of project or activity. Methods for prediction include the kinds of standard procedures and mathematical models used by natural and social scientists and others to forecast the changes likely to occur as a result of a given project or activity. In contrast, methods for evaluation are techniques used in the process of putting a relative value on different impacts and establishing a preference ordering among alternatives. This differentiation between identification, prediction and evaluation makes it possible to distinguish between professional judgments on the nature of expected impacts (identification and prediction) and the kinds of value judgments that are associated with making trade-offs and ranking alternative actions (evaluation).

The first column in Table 2 divides impact identification procedures into four categories. First are checklists, i.e. lists of environmental factors to be considered (or questions to be answered) in analyzing the impacts of a given type of project component. For example, if the project includes an impoundment, the relevant checklist might call for an estimate of the extent of expected change in dissolved oxygen in the reach of stream below the proposed dam. The guidance issued by agencies often includes checklists to assist their field level planners in carrying out environmental assessments. A second category of materials consists of matrices (or tables) that array the components of a given type of project (e.g. dredging, spoil disposal) against the characteristics of the environment that may be affected by these components (e.g. dissolved oxygen, benthic organisms). A dot is indicated in cells of the matrix for which there is a postulated relationship between a project component and an environmental

*Table 2. Environmental assessment methods**

Impact identification	Impact prediction	Impact evaluation
Checklists	Single discipline procedures	Environmental evaluation procedures
Factors to consider	Air, water quality models	Judgments by panels or interdisciplinary teams
Questions to answer	Techniques for visual impact analysis	Weighted average of factors
Matrices	Noise forecasting techniques	
Network diagrams	Social science forecasting methods	Multi-objective evaluation procedures[c]
State-of-the-art reviews[a]	Biological science forecasting methods	Mathematical programming
		Statistical decision analysis
	Cross impact procedures[b]	
	KSIM	
	Systems dynamics models	
	DELPHI panels	

* Except where otherwise indicated, a discussion of the entries listed in the body of the table is given by Canter.[7]

a See, for example, the article by Hagan and Roberts.[5]

b A general discussion of these procedures is given by Sage,[9] and a discussion of application of these procedures in water resources planning is given by Mitchell *et al.*[10]

c An overview of these approaches is provided in Cochrane and Zeleny,[11] and applications in water resources are reviewed by Cohen and Marks.[12]

characteristic; sometimes numerical values are used to indicate the "strength" of this relationship. The third category of impact identification materials consists of network diagrams. These are ordered collections of boxes and arrows that are used to indicate the types of cause–effect relations that may be set in motion if a particular type of project is implemented; e.g. an impoundment may lead to thermal stratification which in turn causes a shift in dissolved oxygen concentration, etc. The fourth category of materials consists of reviews of the literature on impacts associated with a given type of project; the several reviews mentioned in the previous section (e.g. Hagan and Roberts[5]) illustrate this category of materials.

The second column in Table 2 concerns procedures for environmental impact prediction. These can be divided into two broad categories: single discipline procedures and cross impact methods. The former typically provide in-depth treatment of a small group of related factors and constitute the well established products of traditional research. This single discipline category can be described by examples: techniques used by sanitary engineers to predict water quality changes caused by the impoundment of free flowing streams; approaches developed by landscape architects to describe the visual impacts of water resources projects, and procedures used by biologists to estimate the effects of channel modification on fishery resources.

The cross impact methods represent attempts to account for the complete range of factors relevant to a particular forecasting problem when the underlying interrelationships are either too diverse or too poorly understood to be treated by single discipline procedures. An illustration of a situation that could require a cross impact method is the problem of forecasting changes in land use induced by a project providing flood control and recreation facilities. The variables affecting land use are wide ranging and the relationships between water project outputs and land use are not well understood. Table 2 lists three examples of cross impact methods that have been used recently in the context of water planning. Two of these (KSIM and systems dynamics models) involve the use of mathematical simulation modeling, and the third (DELPHI panels) is a procedure for utilizing the opinions of experts in making forecasts. Details of these applications and a discussion of other techniques in this cross impact category are given by Sage[9] and Mitchel *et al.*[10]

The third column in Table 2 includes two categories of procedures for impact evaluation. The first category consists of methods that have been devised by those concerned primarily with environmental assessments (as opposed to economic assessments); these methods indicate how the results from environmental impact analyses can be used to assist decision makers in ranking alternative projects. These methods typically rely heavily on the judgments of those carrying out the planning. One often noted approach, the "Leopold matrix", involves a matrix of the type described above in connection with impact identification procedures. In this case, however, the matrix is used in the context of a specific project, and the cells of the matrix contain two numerical ratings indicating the magnitude and significance of the interaction between the project component and the environmental condition associated with the cell. Other approaches in this category rely on the development of a single overall measure of a project's worth, as follows: First, all of the important factors (or indicators) that may be affected by the alternative projects are set out; this includes economic and engineering factors as well as environmental factors. For any one alternative, each of these factors is given a numerical score which in some sense reflects the extent of the project's impact in terms of the factor. Weights (i.e. measures of the relative value or significance of the different factors) are then ascribed to each factor and a weighted average of factors is computed; it serves as an index of the overall value of the alternative. Weighted averages are computed for each alternative and used to aid in the selection of a proposed action. A discussion and critique of typical applications of this approach to incorporating environmental assessments in water resources planning is given by Ortolano.[13]

The second category in the third column includes methods that have been devised by those concerned with the evaluation of alternatives in the face of multiple objectives. Although environmental quality may be included as one of the objectives, those devising such methods typically have a much more general orientation and are not preoccupied with environmental assessment *per se*. The general literature on this subject includes such topics as mathematical programming and statistical decision analysis and is reviewed in the works edited by Cochrane and Zeleny[11] and Zeleny.[14] The subset of the literature that concerns water resources planning has been reviewed by Cohen and Marks;[12] additional relevant materials are contained in Haimes *et al.*[15]

One of the issues that preoccupied many researchers in the early 1970s was whether or not a single, general-purpose environmental assessment method could be developed to meet the requirements for environmental assessments imposed by the National Environmental Policy act of 1969. In considering this question, the Stanford Workshop on the Environmental Impacts of Water Projects concluded that a single, general-purpose environmental assessment method was an impractical goal and not one that they would choose to pursue.[6] They preferred to leave aspects of methodology development to the numerous researchers in a variety of well established disciplines who had, for generations, been pursuing questions relating to forecasting the effects of water projects and evaluating alternative water resources proposals.

For the members of the Stanford Workshop, the key issues in ensuring that environmental factors received adequate consideration in water resources planning did not relate to environmental assessment methods *per se*; rather, the key issues concerned the ways in which the results from these environmental assessments were being (and could be) utilized in water resources planning and decision making. These issues are pursued in the remainder of this paper. The next section concerns results from studies documenting the extent to which the environmental assessments carried out in response to NEPA have influenced federal water resources planning and decision making. The section following it concerns ways in which the information generated as a consequence of environmental assessments can be integrated more effectively into processes for water resources planning and decision making.

INFLUENCE OF ENVIRONMENTAL ASSESSMENTS ON PLANNING OUTCOMES

In 1973 a series of research studies was initiated at Stanford University to determine the extent to which various federal water resources agencies were integrating environmental considerations into their planning and decision making in response to NEPA. Of particular concern was the field level implementation of the "environmental assessment process" set up by Section 102(2)(C) of NEPA and by the associated guidance issued by the President's Council on Environmental Quality.[16] This process requires a federal agency proposing an action that may have a significant impact on the environment to prepare a draft environmental impact statement (EIS). The draft EIS is to contain an environmental assessment of the proposed action and alternatives to it, and this draft is to be circulated for review and comment by other agencies and various segments of the public (e.g. citizens' groups). After the draft EIS has been circulated, the agency proposing the action must respond to any comments it receives by, at the very least, modifying the EIS. Other, more substantive responses to these comments include: the addition of so-called "mitigation features" (i.e. project components designed to offset adverse effects); the shift to a different action; or the decision not to proceed with any action. After modifying the draft EIS, a final EIS must be circulated before the agency can implement the recommended action.

Detailed case studies

Carmel River case study — As part of the above-noted research effort, Randolph and Ortolano carried out two detailed case studies of Corps of Engineers planning in Northern California. One of the case studies involved "pre-authorization planning" on the Carmel River in California.[17] Pre-authorization planning is preliminary in nature and generally leads to a recommendation for a specific action by the Corps of Engineers to Congress. The Carmel River investigation was initiated by the San Francisco District after the passage of NEPA, and thus it provided an opportunity to gauge the influence of NEPA on early planning decisions, especially decisions relating to the initial formulation and ranking of alternatives.

The Carmel River case study demonstrated that the attitudes of persons responsible for managing a planning study can play a key role in determining the extent to which environmental factors are considered (cf. White[18]). During the early stages of planning, the process of conceiving and formulating alternatives was dominated and controlled by the "study manager", a member of the San Francisco District's Planning Branch; the study manager focused on several alternative multi-purpose reservoir projects for dealing with flooding and water supply problems. The "environmental coordinator", the member of the District's Environmental Branch responsible for directing the environmental assessments, was unable to use environmental factors to broaden the range of alternatives. The one place where environmental assessments had a significant influence on decision making was in connection with the action that emerged as the one to be recommended. In this case, the detailed assessments conducted in preparing the draft EIS led to the introduction of various mitigation features in the project design (e.g. inclusion of provisions for a fish hatchery). It is noteworthy that the portion of the Principles and Standards calling for a plan emphasizing an environmental quality objective played a much more significant role than NEPA requirements in broadening the range of alternatives and in fostering substantive coordination between the environmental specialists and the study manager.

New Melones case study — A second case study concerned "post-authorization" planning, i.e. the detailed engineering and design studies carried out after Congressional authorization of a project. The particular study examined was the Sacramento District's planning for the New Melones project on the Stanislaus River in California.[19] Much of the planning had taken place prior to NEPA's passage, and the case study was designed to examine NEPA's influence on planning and decision making under these circumstances.

The influence of NEPA was, for the most part, restricted to effects on coordination and on the mitigation of adverse effects. With regard to coordination, the case study demonstrated that the process of review and comment on various NEPA related documents (e.g. the draft EIS) can be an effective means of generating useful information from other agencies and citizen's groups. In part because of limited distribution, this review and comment process was ineffective in soliciting information from citizens who were not affiliated with groups (cf. Hill and Ortolano[20]). With regard to mitigation, the information generated for preparation of various environmental assessment documents contributed to the introduction of the following project features to offset adverse environmental impacts: (1) a plan to preserve fish and riparian wildlife habitat areas to offset the areas of such habitat that were to be inundated; (2) the preservation of a 4 mile reach of stream suitable for recreational kayaking to partially offset the loss of a popular "white water" recreation area upstream of the New Melones dam; and (3) the purchase of land containing cave resources that would offset the inundation of what the National Speleological Society considered to be valuable cave resources.

The New Melones case study demonstrates the great difficulties involved in attempting to force an agency to modify its position in response to environmental concerns, when these concerns are made known very late in the planning process. The late stage opposition to the New Melones project was substantial: a law suit was filed, court injunctions were used to halt construction, supplemental environmental studies were ordered by the courts, and the citizens of California actually voted on a project related issue that was included as a proposition in a statewide election. Despite all this, there were no major changes in the project as it was conceived before the opposition began. As elaborated by Randolph and Ortolano,[19] there were significant institutional factors (e.g. agreements made with the US Bureau of Reclamation, financial commitments made to the project as designed) that constrained the Corps' ability to initiate a major re-analysis and a reiteration of their planning process.

Mail questionnaire surveys

Another aspect of the research on how NEPA has influenced federal water resources planning involved the use of mailed questionnaries administered to field level water resources planners in the Corps of Engineers and the Soil Conservation Service (SCS). Three different questionnaires were used, one for each of three different types of field level planning personnel in the District Offices of the Corps and the State Offices of SCS (i.e. planning supervisors, study managers and environmental specialists). Questionnaires were mailed out to each Corps' District Office and each SCS state office early in 1974. SCS returned a total of 139 completed questionnaires (99% response rate), while 103 were returned from the Corps (93% response rate). Complete details regarding the methodological aspects of the survey are given by Hill and Ortolano,[21] and a discussion of all aspects of the survey results is given by Hill.[22] Although the surveys yielded information on a wide range of topics, only a few of those topics will be noted herein.

Formulation of alternatives – One issue explored with the mail survey concerned the influence of environmental assessments on the formulation of alternatives. This was examined by asking respondents to answer several "project specific questions", i.e. questions that referred to a planning study that each respondent had been involved with recently. (These were "pre-authorization planning" studies for Corps respondents and the equivalent for SCS respondents.) One such question asked if any alternative actions or project modifications had been suggested as a result of environmental assessments done for the particular planning study that they were using to answer the project specific questions. Approximately half of the respondents indicated that environmental assessments had served this function. A second part of the question asked respondents who had responded in the affirmative, to indicate the nature of these alternatives or modifications. Most of the responses here could be categorized as either design modification (e.g. eliminating some channel modifications, reducing the level of flood protection) or fish and wildlife mitigation features (e.g. maintaining a minimum flow below a dam to protect fishery resources). Only three respondents in each agency indicted that a non-structural measure (e.g. flood plain zoning) was suggested, and virtually none of the respondents mentioned the so-called "no-project alternative" as a suggestion. Thus, while there were suggestions for new alternatives in roughly half the cases, most of the suggestions involved the types of features traditionally considered in project planning; non-structural approaches to dealing with water problems were rarely mentioned. It is significant that roughly half of the assessments did not lead to any suggestions regarding new alternatives. Taken together, these results suggest that environmental assessments were not being used to broaden the range of alternatives considered in planning by the agencies.

Evaluation of alternatives – Another issue explored with the mail survey concerned the influence of environmental assessments on the evaluation or ranking of alternative projects. This was examined by asking respondents whether any alternatives had been eliminated from further consideration on the basis of environmental assessments. In this case, only one quarter of the respondents answered in the affirmative. Those responding positively were asked to indicate the nature of the alternative eliminated. In nearly all such cases, the alternatives that were eliminated involved channel modification works. This may well reflect the high level of controversy surrounding channel modifications in the early 1970s. (See, e.g. the US House of Representatives hearings on this subject[23].) In any event, the results to this question do not suggest that environmental assessments played a significant role in eliminating alternatives. It could be, of course, that there were very few alternatives that should have been eliminated because of adverse environmental effects. The mail survey data cannot be used to clarify this point.

As Hill and Ortolano[24] point out, the overall conclusions to be drawn from such data depends very much on one's expectation regarding what NEPA was to accomplish. Based on their expectation that NEPA was to force federal agencies to consider environmental factors equally with engineering and economic factors in planning and decision making, Hill and Ortolano interpret the data as indicating that NEPA had not been very effective. Their more complete set of data indicate that, in the Corps and SCS planning studies underway in early 1974, NEPA had not greatly affected either the types of alternatives being considered or who and what influenced the formulation and evaluation of these alternatives.[24]

Organization design studies

One of the findings to emerge from the studies conducted by Hill, Randolph and Ortolano was that organization design seemed to play a significant role in determining the extent to which environmental factors are integrated into agency planning and decision making. Jenkins[25] pursued this subject by analyzing the designs of two water resources planning offices: the San Francisco District of the Corps of Engineers and the Santa Clara Valley Water District. The discussion below is restricted to the Corps District Office since it is of more general interest. Although much of what Jenkins did (e.g. testing hypotheses of the contingency theory perspective of organizations) is not germane to this paper, there is one aspect of his study that is especially relevant. It concerns questions relating to the extent to which different individuals and groups in a flood control planning study influence the study outcomes.

Jenkins carried out much of his data gathering by interviewing the planners and environmental specialists in the offices included in his investigation. In the case of the San Francisco District Office of the Corps, this involved interviews with the ten individuals in the District who were involved significantly in flood control planning at the time of his interviews (1975).

To examine who influenced the outcome of flood control planning studies, Jenkins divided the tasks in water planning into four categories: problem definition, alternative

Table 3. Amounts of influence of key participants in planning decisions: San Francisco District Office – Corps of Engineers, 1975

Decision making task	Amount of influence	
Problem definition	Local interests	4.1
	Superiors	3.7
	Planners	3.7
	Outside agencies	3.6
	Environmental groups	3.1
	Environmental specialists	3.1
	General public	2.5
Alternative formulation	Planners	3.9
	Local interests	3.7
	Superiors	3.5
	Environmental groups	3.3
	Environmental specialists	3.1
	Outside agencies	2.8
	General public	2.6
Impact assessment	Environmental specialists	4.5
	Outside agencies	3.5
	Environmental groups	3.5
	Planners	3.1
	Superiors	2.8
	Local interests	2.8
	General public	2.7
Plan selection	Superiors	4.0
	Local interests	3.9
	Planners	3.6
	Environmental groups	3.2
	Environmental specialists	3.0
	General public	3.0
	Outside agencies	2.8

Source: Adapted from Jenkins,[25] p. 277.

formulation, impact assessment and plan selection. Those interviewed were asked to consider the influence of the following groups: engineering planners, environmental specialists, supervisors and other superiors, the general public, environmental groups, and local interests. Each interviewee was asked to indicate the amount of influence of the aforementioned groups on a scale from 1 (little or none) to 5 (a very great deal). The results, shown in Table 3, indicate that the predominant influence of environmental specialists and groups is in the task of impact assessment. This reflects one of the major preoccupations of such specialists, namely, the preparation of environmental impact statements. Moreover, many of these specialists were under pressure to prepare EIS's for projects that were either under construction or part of operation and maintenance programs. Jenkins observed:

> Several respondents referred to the environmental unit as an 'EIS factory'. The pressure to prepare impact statements to meet legal requirements has kept environmental specialists from becoming very involved as members of project teams for planning studies.

Taken together, the results from all of the research studies referred to in this section indicate that the influence of environmental assessments on the outcome of water resources planning studies has not been great, and that there is a notable lack of integration of the results of assessments into other aspects of the planning process. (These observations are consistent with those made by White[18,26] in the context of large-scale water resources developments in Africa.) This lack of integration has been recognized by agencies like the Corps of Engineers, and efforts have been made to modify the ways in which water resources planning studies are carried out. The discussion below makes note of some of the research that has been used to guide changes in the Corps' planning process that are now being implemented.

INTEGRATING ENVIRONMENTAL ASSESSMENTS INTO WATER PLANNING

As indicated above, environmental assessments are very often divorced from other planning activities; they have often been conducted to meet procedural reporting requirements without being viewed as an integral part of planning. In 1972, a group of researchers in the Civil Engineering Department at Stanford, working in collaboration with the US Army Engineers Institute for Water Resources, began to think of ways to effectively link environmental assessment activities with more traditional planning activities. The result was a planning process, soonafter labeled the "iterative, open planning process" (IOPP), that was proposed for use in the District Offices of the Corps of Engineers.

Iterative, open planning process

The IOPP, which has been described elsewhere by Ortolano,[27] is based on a few simple concepts that have far reaching implications for the way in which water resources planning is carried out. The IOPP is "open" in the sense that it relies on continual two-way communication between agency planners and a wide range of interested citizens and government agencies beginning at the earliest stages of a planning study. The IOPP is iterative in that it calls for the concurrent (as opposed to sequential) performance of the four traditional planning activities: problem definition, formulation of alternatives, impact assessment and plan ranking. (For a description of water resources planning that relies on the sequential performance of these tasks, see Mussivand.[28]) At any point in the process, information from each of the four planning activities influences each of the other activities. For example, the assessment of impacts may reveal new concerns of affected citizens. Thus, the information from the impact assessment activity "feeds back" to the problem definition activity, which in turn, may be expected to influence the alternatives that are considered.

The need for an open process is based on the premise that the "public interest", which is mandated as the basis for decision making by federal water agencies like the Corps of Engineers, cannot be determined through objective analysis performed by technical specialists.[29]

Rather, the concept of the public interest is more appropriately conceived of in terms of a process whose direction and outcome can be influenced by the various agencies, citizens' groups, and individuals that may be affected. Other arguments commonly made by advocates of an open planning process are: other agencies and various elements of the public have relevant expertise and information to contribute to water agency planners; public participation is an important part of the process of placing relative values on environmental effects (planners cannot do this alone because no objective techniques exist for doing so); late stage opposition to proposals can be minimized by involving potential opponents in planning from the outset; and a host of recent laws and regulations require significant levels of public participation in water resources planning. (For other and often opposite perspectives on the appropriateness of public involvement activities, see Pierce and Doerksen.[30])

An iterative process is required if continual public involvement is to be encouraged and accommodated. As planning proceeds and information is provided to other agencies and various segments of the public, new concerns and problems may become evident. These newly delineated concerns and problems may call for the abandonment of previously favored alternatives, the formulation of new alternatives, the assessment of impacts that had not been previously considered, etc. With a rigid, sequential process, these types of activities are often viewed as setbacks and not as important opportunities to be responsive to the needs of the various affected interests.

There are several ways in which the IOPP serves to integrate environmental assessments into planning. For one thing, it makes no distinction between environmental impacts and economic impacts; all impacts that are relevant to decision making need to be analyzed and evaluated, regardless of their placement in one taxonomic category or another. Also, because of its deliberately iterative nature, the IOPP requires that information regarding the impacts of alternative actions be considered from the earliest stages of planning. In addition, the information from continual public involvement and inter-agency coordination can help those engaged in impact assessment to decide on which impacts to analyze in detail.

Another way in which the IOPP facilitates the integration of environmental assessments into all other planning activities is provided by the notion of "evaluative factors", i.e. "the goals, concerns, constraints, etc. that various decision makers and effected publics consider important in ranking alternative actions".[27] These evaluative factors are established on the basis of: the judgments of planners and technical specialists within the water agency, the requirements imposed by relevant laws and regulations, and the concerns of interested citizens and administrative agencies other than the water agency. Evaluative factors provide the basis for determining the impacts that need to be assessed in detail, and they play a central role in problem definition, the formulation of alternatives and plan ranking. As such, they serve to drive the entire planning process and link all four planning activities together.

The IOPP has gone beyond the stage of being an exercise in the articulation of concepts. It has, in many respects, been adopted by the Corps of Engineers in their regulations for implementing the Water Resources Councils' mandate for multi-objective planning.[31] In addition, as part of the overall effort to implement these regulations, the IOPP was subjected to a formal testing and evaluation process in a real-world planning context. Some of the results from this field test indicate potential problems in implementation, and these are noted below.

Potential problems in implementing the IOPP

The IOPP was field tested in the context of a study of flooding on San Pedro Creek in

Pacifica, California; the study, carried out between 1973 and 1975, was largely conducted by the San Francisco District Office of the Corps of Engineers. An account of the field test of the IOPP has been given by Wagner and Ortolano.[32]

The three results from the field test that are especially relevant to issues involving the integration of environmental assessments into other activities carried out as part of a planning study concern: (1) the study manager's lack of authority to control the timing of study activities; (2) difficulties involved in effecting early, substantive coordination with other agencies; and (3) problems associated with the use of interdisciplinary planning teams to coordinate the activities of technical specialists.

The first of these results concerns the ways in which various study activities are linked over time. The IOPP requires that information from environmental assessments be made available to decision makers before tentative conclusions are reached regarding which alternative to recommend. This information was not made available in a timely fashion in the field test, and the reasons can be traced to the limited authority of the study manager to control the timetable of the study. Although a Corps of Engineers study manager typically has great influence over the direction of a study (e.g. which alternatives are examined), he must go through formal channels in order to have study-related work performed by technical specialists in other organizational units. For example, before members of the Environmental Branch in the San Francisco District Office can conduct an environmental assessment for a study manager, a formal written request for the assessment must travel through channels to the Chief of the Environmental Branch. Upon receiving the request, the Environmental Branch Chief assigns the work to one or more of his staff members and indicates the time at which the work is to be completed. The study manager can exert informal pressure to have the work completed to meet his own scheduling requirements, but he has no authority to force his own scheduling priorities on the Environmental Branch Chief. The study manager's limited ability to control the timing of studies carried out by specialists can make it difficult to meet the coordination requirements associated with the IOPP.

The second of the field test results concerns coordination between Corps planners and staff members in other agencies. In commenting on the field test, Wagner and Ortolano[32] noted the advantages of doing more early, informal coordination than was actually accomplished. Personnel in other agencies indicated that they would have welcomed early involvement in planning and that this early involvement would have been useful. There is evidence to suggest that Corps planners frequently do not effect substantive inter-agency coordination early in their planning studies.[33] To the extent that such coordination is not carried out, the IOPP will be less than fully effective.

The third result concerns the use of interdisciplinary teams to coordinate the efforts of the various technical specialists involved in a study. Such teams are called for by the Corps of Engineers regulations implementing the Principles and Standards, and they could be useful in the context of the IOPP; for these reasons an interdisciplinary planning team was used in the field test. In evaluating the field test some of the District personnel indicated dissatisfaction with the effectiveness of the team leadership and complained that their time was not used efficiently during team meetings. These problems point to the need for extensive training of team members, especially team leaders, in various aspects of group decision processes.[9] The field test also indicated that the interdisciplinary planning team concept requires that members possess an identification with and a commitment to the team. These cannot develop unless the team concept is supported by the relevant branch and section chiefs within a District's hierarchy and unless staff members within these branches and sections are given adequate time to work within the context of the planning team. Both the general literature on the use of interdisciplinary groups (e.g. Galbraith[34]) and the more specialized literature on the use of such groups in water resources planning (e.g. Flack[35]) indicate that there are a host of factors like the ones noted above that have to be dealt with if planning teams are to be an effective organizational arrangement for implementing the IOPP.

Because of its similarity to the IOPP, the planning process recently adopted by the Corps of Engineers will likely lead to similar problems. Despite these problems, the Corps' new planning process should encourage an increase in the consideration given to environmental factors in *all* planning activities: problem definition, formulation of alternatives, impact assessment and plan ranking. This will be in contrast to the early 1970s during which time environmental factors were considered in the context of impact assessment, but seemed to play a much less significant role in the other planning activities.

CONCLUDING REMARKS

This paper has argued that the principal issues involved in the environmental assessment of water projects do not concern assessment methods *per se*, but the way in which the information developed from the application of various methods is utilized in planning and decision making. The various studies by Hill, Jenkins, Randolph and Ortolano demonstrate that much of the effort in conducting environmental assessments for federal water projects in the early 1970s was not being carefully linked with planning activities relating to the formulation and ranking of alternative actions. Rather, much of the effort was directed at producing environmental impact statements that often seemed to have little influence on the decisions reached in planning studies.

For those who feel that the results of environmental assessments are intended to influence all aspects of planning, including the activities carried out in the early stages of planning (e.g. the initial formulation of alternatives), there are grounds for optimism. A number of federal water agencies have made revisions in their planning procedures that encourage the consideration of environmental factors in all aspects of planning; the Corps of Engineers adoption of a process that is similar to the IOPP provides and example of this. Moreover, the portion of the Principles and Standards that requires the delineation of a plan emphasizing an environmental quality objective should significantly increase the extent to which environmental factors influence initial efforts to formulate alternative actions.

The 1970s represent a period of transition with regard to the way environmental factors are considered in water planning. As a result of NEPA and the Principles and Standards, and the increased levels of public participation in planning, field level water resources planners have been forced to give much more consideration to environmental factors then they have in the past. In some instances, this consideration has been superficial and responsive only to legal requirements to provide an environmental impact statement. In other situations, thorough consideration has been given to environmental factors in all planning activities. At this point, it is impossible to say whether this type of thorough consideration of environmental factors will become a characteristic of water resources planning in the United States.

Acknowledgements – Much of the research reported on herein was conducted as part of the Stanford University research program on the environmental impacts of water projects. This research was funded largely by the US Army Engineers Institute for Water Resources and the Office of Water Research and Technology in the Department of the Interior. This paper relied heavily on the research of the following individuals while they were students in the Department of Civil Engineering at Stanford: William W. Hill, Bryan R. Jenkins, John Randolph and Thomas P. Wagner. It is a pleasure to acknowledge the helpful comments on an earlier draft by Charles M. Brendecke, John Meersman and James E. Price, students in the Department of Civil Engineering at Stanford.

REFERENCES

1. US Water Resources Council, The principles and standards for planning water and related land resources, *Fed. Regist.*, **38** (174) Part III, (1973).
2. California Department of Water Resources, The California State Water Project, *Bulletin No.* 132–66, Sacramento, CA, (1966).
3. Gill, G.S., Gray, E.L. and Seckler, D., The California Water Plan and its critics: a brief review, in Seckler, D. (ed.) *California Water, A Study in Resource Management*, University of California Press, Berkeley, CA, (1971).
4. Goldman, C.R. Ecological implications of reduced freshwater flows on the San Francisco Bay Delta system, in Seckler, D. (ed.) *California Water, A Study in Resource Management*, University of California Press, Berkeley, CA, (1971).
5. Hagan, R.M. and Roberts, E.B. Ecological impacts of water projects in California, *J. Irrig. Drain. Div. Am. Soc. civ. Engrs*, **98** (IRI), 25–48 (1972).
6. Ortolano, L. (ed.), *Analyzing the environmental impacts of water projects*, Report 73–3, US Army Engineers Institute for Water Resources, Fort Belvoir, VA, (1973).
7. Canter, L.W. *Environmental Impact Assessment*, McGraw-Hill, New York, (1977).
8. Dickert, T.G. Methods for environmental impact assessment: a comparison, in Dickert, T.G. and Domeny, K.R. *Environmental Impact Assessment: Guidelines and Commentary*, University Extension, University of California, Berkeley, CA, (1974).
9. Sage, A.P. *Methodology for Large-Scale Systems*, McGraw-Hill, New York, (1977).
10. Mitchell, A. *et al. Handbook of forecasting techniques*, report prepared for the US Army Corps of Engineers Institute for Water Resources, Fort Belvoir, VA, (1975).
11. Cochrane, J. and Zeleny, M. (eds.), *Multiple Criteria Decision Making*, University of South Carolina Press, Columbia, SC, (1973).
12. Cohen, J.L. and Marks, D.H. A review and evaluation of multiobjective programming techniques, *Water Resour. Res.*, **11** (2), 208–220 (1975).
13. Ortolano, L. Water plan ranking and the public interest, *J. Wat. Resour. Plann. Management Div. Am. Soc. civ. Engrs*, **102** (WR1), 35–48 (1976).
14. Zeleny, M. (ed.), *Multiple Criteria Decision Making, Kyoto, 1975*, Lecture Notes in Economics and Mathematical Systems – 123, Springer-Verlag, Berlin, (1975).
15. Haimes, Y.Y., Hall, W.A. and Freedman, H.T. *Multiobjective Optimization in Water Resources Systems*, Elsevier, Amsterdam, (1975).
16. Council on Environmental Quality, Preparation of environmental impact statements: guidelines, *Fed. Regist.*, **38** (147), Part II, 20550–20562 (1973).
17. Randolph, J. and Ortolano, L. NEPA and the considerations of alternatives: a case study of Corps of Engineers planning for Carmel River, *Environ. Aff.* **5** (2) 213–253 (1976).
18. White, G.F. Interdisciplinary studies of large reservoirs in Africa, in Flack, J.E. (ed.), *Proc. of the Conference on Interdisciplinary Analysis of Water Resource Systems*, American Society of Civil Engineers, New York, (1975).
19. Randolph, J. and Ortolano, L. Effect of NEPA on the Corps of Engineers' New Melones Project, *Columbia J. Environ. Law*, **1** (2), 233–273 (1975).

20. Hill, W.W. and Ortolano, L. Effects on NEPA's review and comment process on water resources planning: results of a survey of planners in the Corps of Engineers and Soil Conservation Service, *Water Resour. Res.* **12** (6), 1093–1100 (1976).
21. Hill, W.W. and Ortolano, L. *NEPA's influence on federal water planning: part 1, effects of the 'review and comment process' on the Corps of Engineers and the Soil Conservation Service,* EEP Report No. 52, Dept. of Civil Engineering, Stanford University, Stanford, CA, (1975).
22. Hill, W.W. The national environmental policy act and water resources planning in the Corps and SCS: effects and effectiveness, PhD. dissertation, Stanford University, Stanford, CA, (to be published).
23. US House of Representatives, *Stream channelization – what federally financed draglines and bulldozers do to our nation's streams,* House Report No. 93–500, based on hearings before the Conservation and Natural Resources Subcommittee of the Committee on Government Operations, 93rd Congress, 1st Session, Washington, DC, (1973).
24. Hill, W.W. and Ortolano, L. NEPA's effect on the consideration of alternatives: a crucial test, *Nat. Resour. J.*, (to be published).
25. Jenkins, B.R. Changes in water resources planning: an organization theory perspective, PhD. Dissertation, Stanford University, Stanford, CA, (1977).
26. White, G.F. Organizing scientific investigations to deal with environmental impacts, in Farvar, M.T. and Milton, J.P. (eds.), *The Careless Technology,* Natural History Press, Garden City, NY, (1972).
27. Ortolano, L. A process for federal water planning at the field level, *Water Resour. Bull.* **10** (4), 766–778 (1974).
28. Mussivand, T.V. Application of critical path method to water resources planning, *Water Resour. Bull.* **8** (4), 685–696 (1972).
29. Fox, I.K. Policy problems in the field of water resources, in Kneese, A.V. and Smith, S.C. (eds.), *Water Research,* Johns Hopkins Press, Baltimore, MD, (1966).
30. Pierce, J.C. and Doerksen, H.R. *Water Politics and Public Involvement,* Ann Arbor, MI, (1976).
31. US Army, *Planning process: multiobjective planning framework,* ER 1105–2–000, Office of the Chief of Engineers, Washington, DC, (1975).
32. Wagner, T.P. and Ortolano, L. *Testing an iterative open process for water resources planning,* Report 76–2, US Army Engineers Institute for Water Resources, Fort Belvoir, VA, (1976).
33. Randolph, J. Influence of NEPA on Corps of Engineers Water Planning in California, PhD. dissertation, Stanford University, Stanford, CA, (1976).
34. Galbraith, J. *Designing Complex Organizations.* Addison-Wesley, Reading, MA, (1973).
35. Flack, J.E. (ed.), *Proc. of the Conference on Interdisciplinary Analysis of Water Resource Systems,* American Society of Civil Engineers, New York, (1975).

Water Supply & Management, Vol. 2, pp.177-185. 0364-7714/78/0601-0177$02.00/0

Environmental Issues of Large Interregional Water Transfer Projects

GENADY GOLUBEV*

The problem of large water transfer projects is triune and consists of technology, socio-economy, and the environment.[1] Two principal questions exist on the environmental side of the problem:

(1) How does one evaluate efficiency of a project incorporating environmental issues which are sometimes not measureable?

(2) How does one predict environmental consequences of large water transfer projects?

In the planning stage, these questions are necessary for (a) assessment of the project's efficiency; (b) comparison of various options of water transfer schemes; and (c) comparison of various other alternatives with transfer schemes.

At the implementation and operational stage of the project, these questions are needed for optimal control of the environment. One of the obstacles in solving them concerns the problem of complexity and related problems of uncertainty. The number of components and links in a geoecosystem is very high. For instance, a quite simple geoecosystem of pine trees in sandy soils has about 2000 links. A simple set of geoecosystems of taiga forests covering an area of 2 X 4 km has some 20,000 links.[2] Consequently, the larger the territory, the more numerous the components and links. In moving up the hierarchical ladder, some links and components can be disregarded, just as some new ones will arise.

Thus, complexity of a natural environment grows with an expansion of the territory under study and, in the case of IWT, with the increase of its size. The following question arises: How certain can our assessments and forecasts of IWT projects be?

It seems to be obvious that uncertainty is a direct function of complexity, i.e. (1) the number of components and links is high, and our knowledge of all of them is inadequate for practical needs; (2) usually, only the most important components and links are regarded. However, in the taiga example mentioned above, some 50–80% of the information has been lost when out of 160 components, only the most important were taken into account;[2] and (3) uncertainty also increases if a question under study concerns an interaction between various spheres, i.e. atmosphere, hydrosphere, lithosphere, and above all, the biosphere. For example, a forecast of climatic changes – due to an IWT project – at a macroscale requires the modelling of a hydrosphere–atmosphere interaction; the same forecast at a mesoscale level would involve the earth's surface also, that is, the lithosphere and biosphere. The former is difficult, but possible to calculate numerically; the latter is impossible to calculate for the time being.

Let us suppose that we can determine the uncertainty of a project in monetary terms. It will then be an exponential function of a project's size, shown as a solid line in Fig. 1. Expected errors of its determination are shown by dotted lines. When an irrigation project is

* IIASA, Laxenburg, Austria. On leave from the Department of Hydrology, Moscow State University, USSR.

developed for a single field, environmental consequences are known with rather high accuracy. Environmental impacts from a new irrigation system are also predictable but uncertainty would be higher than in the first case and an error of its determination would be greater. Environmental consequences of a set of irrigation systems (such as the California State Water Project) are more difficult to predict, and uncertainty is relatively high. Hence, uncertainty of environmental effects of new, large projects covering subcontinents and continents becomes quite high.

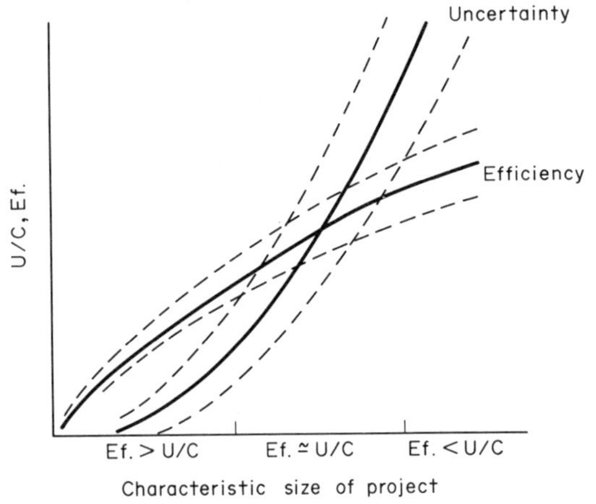

Fig. 1. Conceptual curves of uncertainty and efficiency vs characteristic size of project.

Let us further assume that we can determine efficiency of net benefits of a project – including environmental issues – in monetary terms. Axiomatically, the larger the project, the higher the benefit. The curve of efficiency is also represented in Fig. 1, and takes the intuitive form. Dotted lines show the accuracy of determination of efficiency.

There are three areas in Fig. 1: (1) efficiency is higher than uncertainty; (2) efficiency and uncertainty are the same within their accuracy; and (3) efficiency is inferior to uncertainty. For projects in the first area, the problem of uncertainty is not decisive. As for projects in the second area, they should be postponed unless other, non-economical objectives exist. Projects situated in the third area should be definitely declined, at least for the present.

There is apparently a certain project size above which the uncertainty of environmental consequences is so high that the project is not feasible. Examples of such projects include removal of ice in the Arctic Sea, and construction of huge water reservoirs to unify great river systems of South America, etc. With the progress of science and technology, the curve of uncertainty would shift to the right, and the curve of efficiency to the left; furthermore, location of the three areas would change in such a way that the critical size of a project would increase.

The approach discussed above is purely conceptual. It is, however, possible that very large water transfer projects are situated either in the second or even in the third area of Fig. 1 and, hence, their time for materialization has not yet come.

In the second half of this paper, environmental issues of a large IWT project in the USSR are discussed. There are many projects underway to reallocate water resources throughout the country. They can be grouped into three main options,[3] the second of which is the so-called Integrated Euroasian Option.

According to this option, water should be withdrawn from the Ob Bay near the mouth of the river, and at the first stage, in an amount equal to 25 km^3/yr. The water in Ob Bay is mostly fresh and its volume is comparable with the annual run-off of the Ob River, so that no artificial reservoir is required.

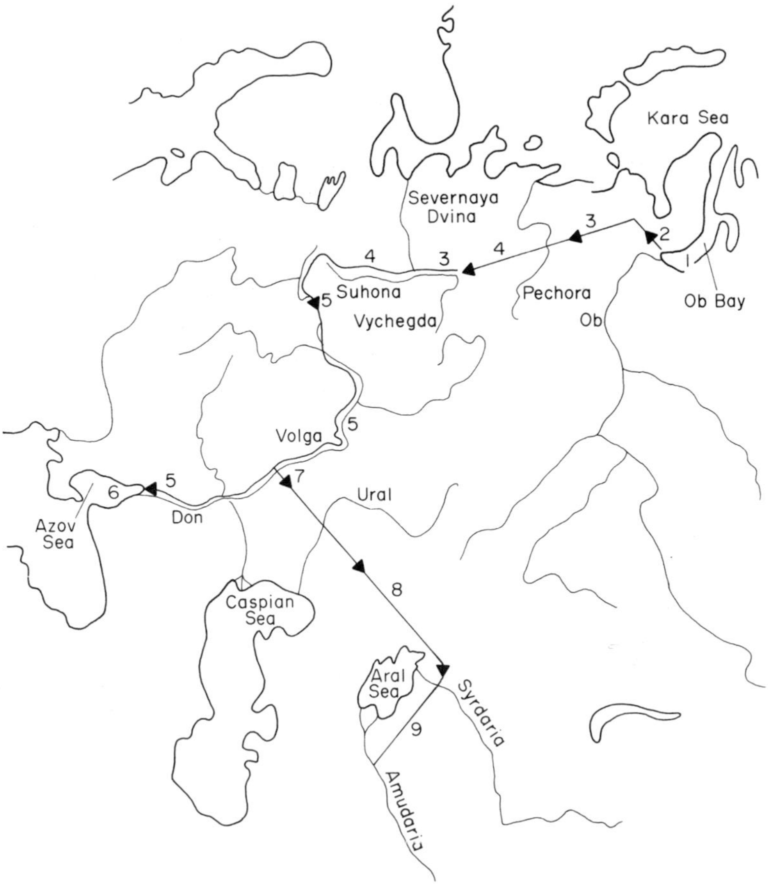

Fig. 2. Division of the IWT project by reaches.

Water would then be pumped over the Ural Mountains to the Pechora River Basin, flowing along the natural channels of the Usa and Pechora. It has been proposed to construct a chain of small reservoirs upstream from the Pizhma River (a left tributary of the Pechora), in order to move water further west. Water in amounts exceeding the withdrawal from Ob Bay would be pumped from the basin and would then use natural river channels of the Vychegda River basin. The same operation would be repeated on the western part of the Severnaya Dvina River basin; that is, by creation of an "anti-river" along the Suhona River by pumping additional water from Severnaya Dvina.

Finally, the water from the Ob, Pechora, and Severnaya Dvina Rivers would reach the Upper Volga, passing through a chain of reservoirs and power plants already created along the Volga River. It would compensate energy spent for pumping on other reaches of the project.

Part of the water in the Lower Volga would go to the Don River, flowing into the Azov Sea in order to improve its hydrologic regime. A major part of the water would travel along the canal between the Volga and Ural Rivers, partially used for new irrigation developments. Then the water would go by canal to the lower reaches of the Syrdaria and Amudaria Rivers for irrigation there. Water currently in use for irrigation of these areas would serve new developments in the middle and upper reaches of both rivers.

Obviously, this is a very complicated project covering enormous territories. My objective has been to determine the environmental consequences of such a big enterprise. In order to decrease uncertainty, the entire project was divided by reaches (see Fig. 2).

A method of trees of consequences or networks has been used. In the country, there are many studies underway on the various environmental aspects of IWT problems. (No reference to them is made here, but the reader may refer to the bibliography in this volume.[4] In some cases, there were no appropriate data, and I based much of my results on a general knowledge of the region and certain assumptions.) As a result, seven networks were constructed. There are no networks for two reaches, the Ural Mountains and the Ural and Syrdaria/Amudaria Rivers, as the engineering area has not been sufficiently cleared. Only the most important components and links are shown. The lower line of each network represents practical problems arising from environmental issues. All the networks are represented in Figs. 3–9. I will not discuss them in detail, but will just mention some of the highlights.

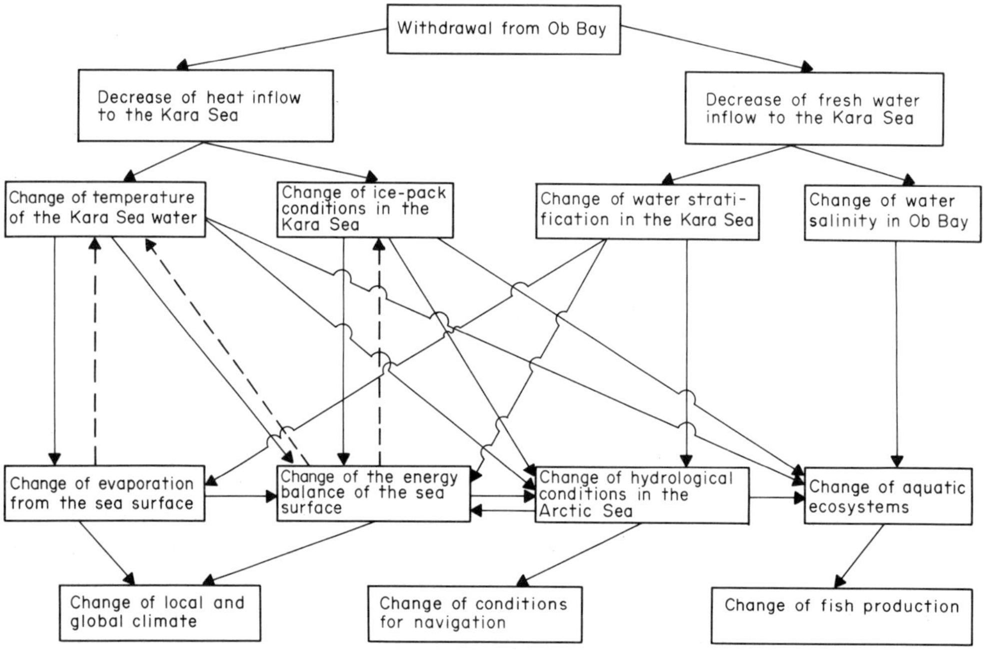

Fig. 3. Network for the 1st reach.

As previously mentioned, it has been proposed to remove 25 km^3/yr from the Ob Bay (Fig. 3). There will certainly be some impacts on many features of the Ob Bay which are very important to predict. At the same time, it is of vital importance to study the impacts on the Arctic Ocean, and a number of institutes in the USSR are working on the problem. The problem is two-fold: (1) the decrease in the amount of fresh water, heat, organic materials and

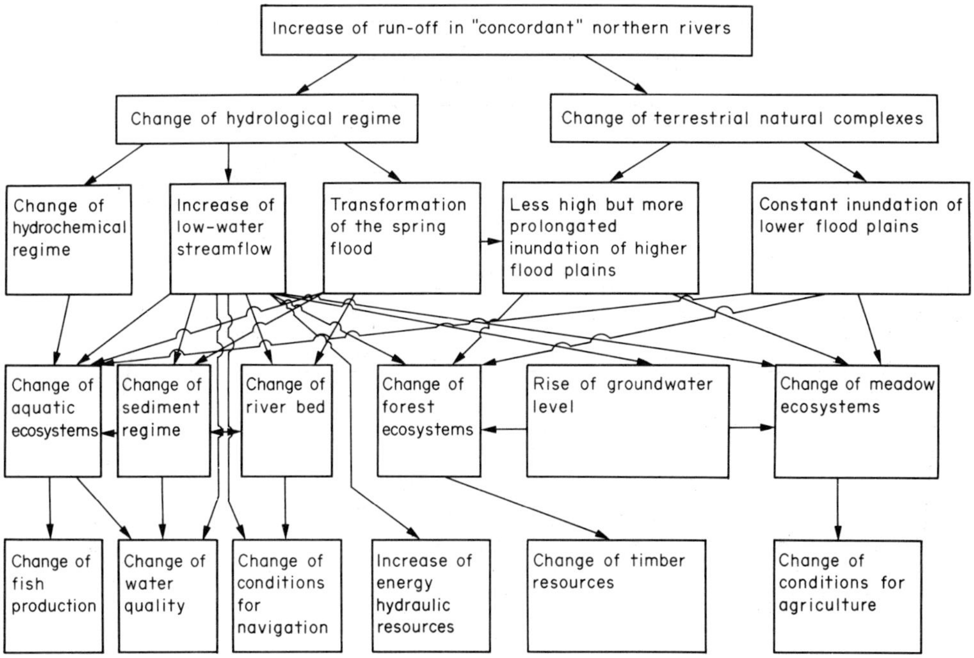

Fig. 4. Network for the 3rd reach.

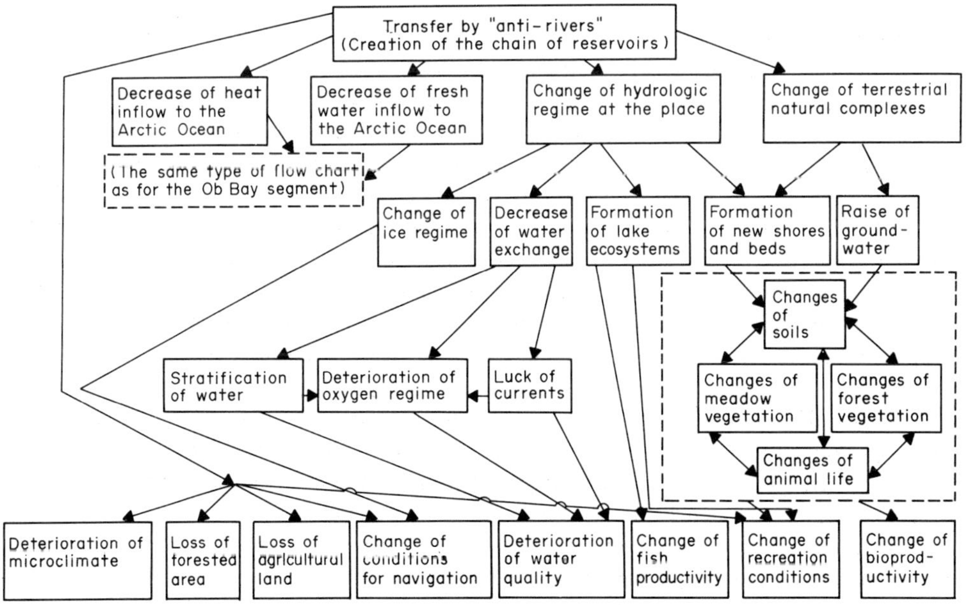

Fig. 5. Network for the 4th reach.

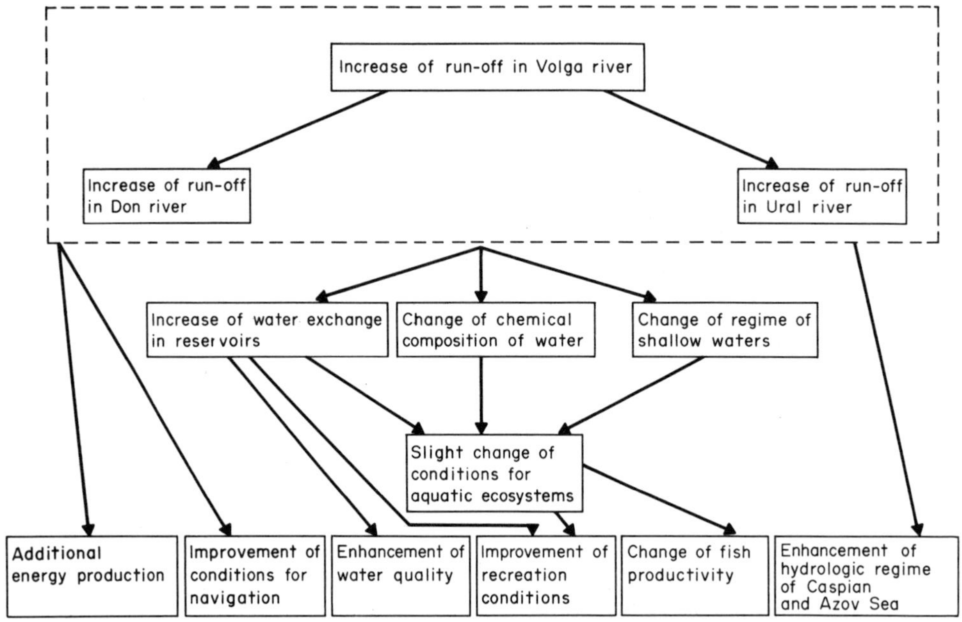

Fig. 6. Network for the 5th reach.

other kinds of inflow; and (2) the change of water stratification and consequent ramifications thereof.

As to the first issue, the mean annual run-off of the Ob River is about 400 km^3/yr. As a result of the project, 6% of this amount would be taken as the first stage and 15% as the final goal. The Kara Sea also receives 620 km^3/yr from the Enisey River. The percentage of withdrawn water in comparison with the run-off of these two great rivers would be correspondingly, 2 and 6%. Eight major rivers flowing from Asia to the Arctic Ocean bring 1950 km^3/yr. Comparing proposed withdrawals with this figure, we would have only about 1 and 3%, respectively. It would seem that all of these values are rather small, and that no considerable impacts are expected; the problem does, however, deserve more precise study.

As to the second issue, there is a stable stratification of water in the Arctic Ocean. Water salinity of 34 parts pro mille or less is situated above water of normal oceanic salinity (35 parts pro mille), and is about 50 m thick. The stratification is stable as less saline water is less dense.

Stable stratification allows for formation of an ice cover over the Arctic Ocean and, more broadly speaking, the actual type of interchange between the Arctic Ocean and the atmosphere with corresponding implications on global climate. A change in the stratification would this lead to many consequences, some of which are unpredictable.

There are two main sources of less saline water: river run-off inflow and net surplus of precipitation over evaporation for the whole area of the Arctic Ocean. Both sources are of the same order of magnitude. A slight decrease in the first component would result in a lesser decrease of the sum of both.

The balance of saline water is dynamic and formed as a result of the previously mentioned components, and the components describing an interchange between the Arctic Ocean and Atlantic and Pacific Oceans. A slight change in one of the components would lead to a different equilibrium of the entire system without drastic changes. However, the question is serious and is now undergoing careful study before a final decision is taken.

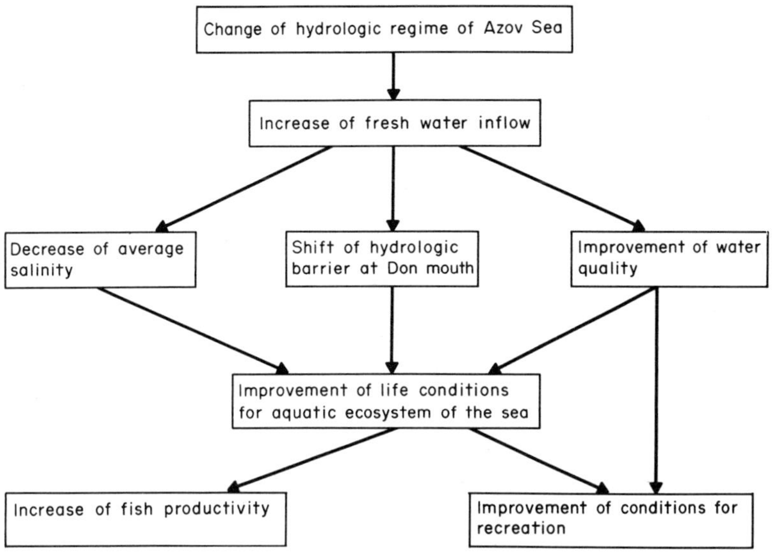

Fig. 7. Network for the 6th reach.

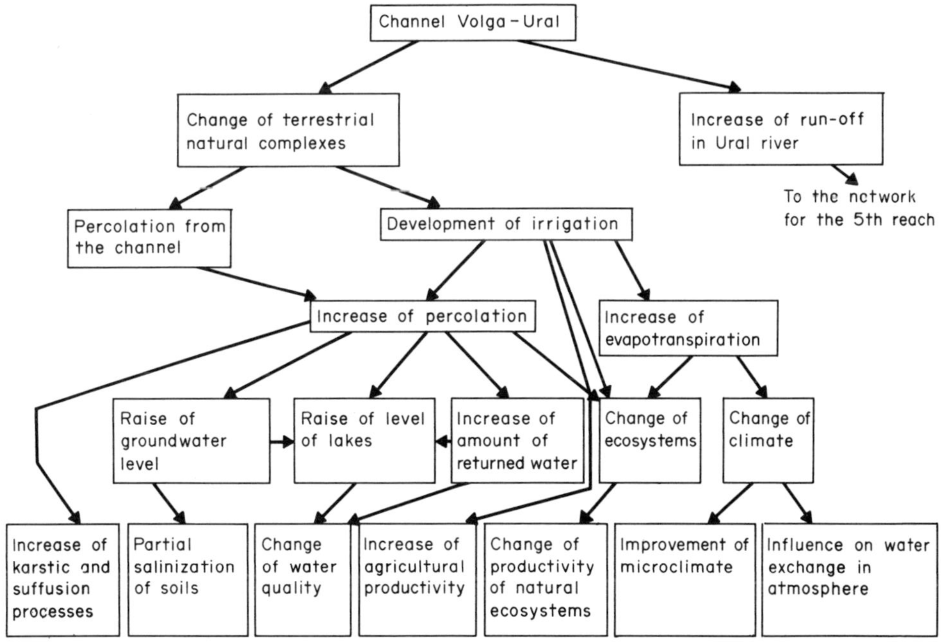

Fig. 8. Network for the 7th reach.

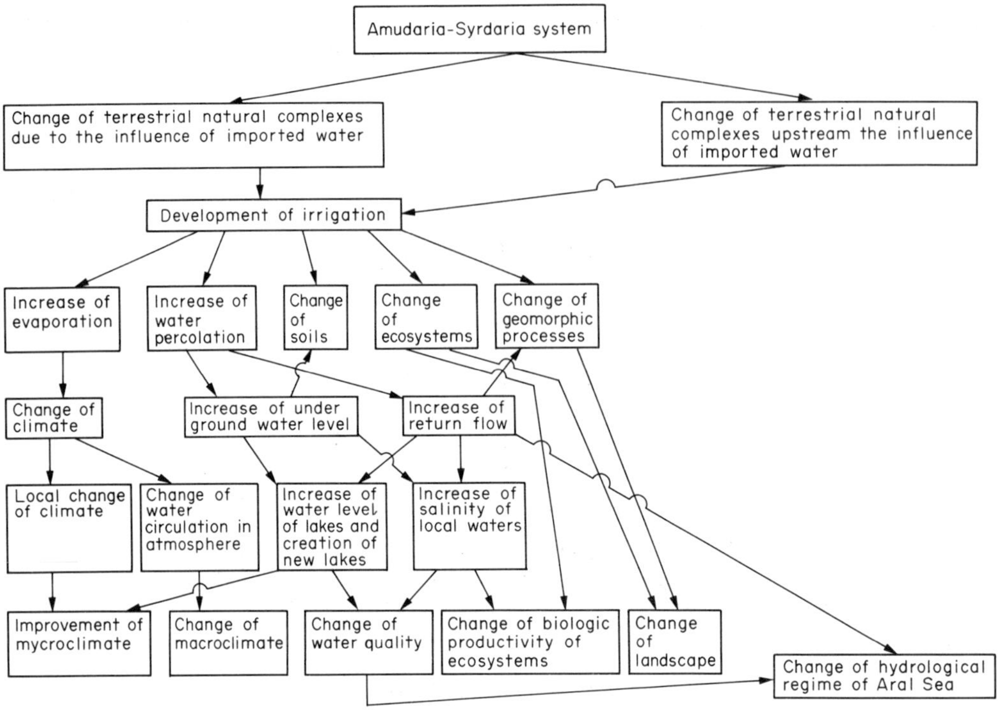

Fig. 9. Network for the 9th reach.

A considerable increase of river run-off will be along the natural channels of the rivers whose direction coincides with that of water transfers. Hydrologic regimes would change noticeably with an ensuing chain of consequences (Fig. 4). At the same time, all systems of interrelations characteristic of the lower levels of the river valleys would be affected drastically. Almost all rural life there is tied to these levels, hence reliable forecasting of events is necessary.

Creation of "anti-rivers" would lead to the classical environmental problem of water reservoirs: loss of land, deterioration of water quality, fish problems, erosion of coasts, etc. (see Fig. 5). The environmental impact of water reservoirs spreads also to adjacent territories. It is estimated[5] that for all water reservoirs of the USSR at the beginning of the 1970s, the area of impact is about 20 or 30 thousand km^2, i.e. approximately 30 or 40% of the total area of water reservoirs. Parallel with the universal problems of water reservoir, there are also regional, or rather, zonal problems. The reservoirs would be in the zone of excess precipitation over potential evaporation. This means that (1) losses of water to evaporation would be roughly the same as before creation of the reservoirs (which is favourable), and (2) additional excess of water would deteriorate terrestrial ecosystems. The main factor here is the rise of the groundwater level, which would bring a number of unfavourable consequences, for example, formation of new swamps and decrease of forest productivity.

An increase of run-off along the Volga River would be between 10 and 25% of the current annual flow at the mouth of the lowest tributary. In fact, it is now a chain of water reservoirs. It would seem that the effect of additional flow would be primarily favourable (Fig. 6).

Additional water transferred to the Azov Sea would bring desirable and positive effects (Fig. 7). The background necessary to carry out this part of the project has been described in this volume.[1]

As to the reach from the Volga to the Ural River, an increase of run-off in the latter is not the goal of the project and the main problems relate to the left branch of Fig. 8. A key problem seems to be percolation of water, either from the main canal which transports water south or, because of development of irrigation instead of dry farming. The percolation process produces a particular chain of consequences, and special attention should be given to proper water management there.

Finally, new water would be brought to the Syrdaria and Amudaria systems for further development of irrigation (Fig. 9). Specific problems of closed basins in arid zones would arise. Development of irrigation means removal of salts from the soils with returned waters. With plans to irrigate about 12 million ha in the year 2000, it will be necessary to remove 2 billion tons of salt. What would be its destination? Part of the salt would possibly go to the Aral Sea, which would turn into a relatively small, very salty lake. Because of the area's topography, not all returned waters would go to the Aral Sea and new closed lakes would form. Two have been formed in the last decade because of new developments in irrigation schemes and one is about 1000 km^2.

The networks (Fig. 3–9) and commentaries are, of course, subjective. The method itself has disadvantages which are discussed in a number of papers. At the same time, however, it is useful to single out the most important phenomenon and to manage and integrate scientific studies of environmental impacts of IWT projects.

REFERENCES

1. Golubev, G.N. and Vasiliev, O.F. *Interregional water transfers as an interdisciplinary problem* (in this issue).
2. Armand, A.D. *Information Models of Natural Complexes,* Nauka, Moscow, (in Russian), (1975).
3. Voropaev, G.V. Problem of reallocation of water resources of the country and its modelling. *Gidrotekh. Melior.*, 9, (in Russian), (1976). See also Voropaev's paper in this issue.
4. Golubev, G.N., Grin, G., Melnikova, G. and Whetstone, G. *Bibliography on water transfers* (in this issue).
5. Vendrov, S.L. and Diakonov, K.N. *Water Reservoirs and Environment,* Nauka, Moscow, (in Russian), (1976).

PART III: Bibliography on Interregional Water Transfers

Water Supply & Management, Vol. 2, pp. 187–211. Pergamon Press, 1978. Printed in Great Britain.

Bibliography on Interregional Water Transfers

GENADY GOLUBEV,* GALINA GRIN,** GALINA MELNIKOVA* and GEORGE WHETSTONE†

A bibliography on Interregional Water Transfers is presented below. Those sources have been selected which were related directly to problems and schemes of interregional (or large-scale) water transfers.

The bibliography contains 502 entries and as a rule, short notes have not been included. The majority of the publications are in languages of the Latin alphabet; the remainder are in the Cyrillic alphabet (Soviet papers), with corresponding translations into English.

As a considerable amount of publications are included in the list, it has been necessary to form sections according to items. Each section is listed alphabetically according to authors. The following sections are listed: General, Regional, Economy, Environment, Hydrology, Technology, Methodology and Water Management, and Politics and Law. The Regional section has been sub-divided into continents, plus the USSR which is situated in two continents. Of course, the divisions are quite relative. Many publications can be related to two or even more sections. We had intended to mention each source only once, but in about twenty instances the source is included in two or more sections.

The preparation for the bibliography has been carried out by an international group through IIASA's coordination. Professor G. Whetstone of the USA has kindly supplied IIASA with his voluminous bibliography on water transfers, which was one of the main sources of the present list. Selection from Professor Whetstone's bibliography has been made by the co-authors from IIASA. Dr. G. Grin, USSR, has prepared the main part of the Soviet bibliography. Final selection of sources, their divisional allocation and preparation for publication has been done by Professor G. Golubev and Dr. G. Melnikova, IIASA.

GENERAL

Agnew, Allen F. and Schuy, David F. *A new look at interbasin transfer,* Wenatchee Conf., Mimeo., 13pp. (April, 1972).

Bagley, Jay M. Water importation into arid lands, *Science,* **162,** 478–479 (1968).

Bagley, Jay M. and Smiley, Terah L. (eds), Water importation into arid lands (Symposium AAAS), in McGinnes and Goldman, *Arid Lands in Perspective,* Univ. of Arizona Press, (1968) pp. 337–421.

Banks, Harvey O. Teamwork in the solution of water problems, *Proc. Am. Soc. civ. Engrs,* **84** (IR-1), Paper 1497, 7 pp., DISC: (IR-3). Paper 1784, 23–25; **85** (IR-1), 67 (1958).

*IIASA, Laxenburg, 2361 Austria.

**Institute of Water Problems, USSR Academy of Sciences, Moscow, USSR.

†Texas Tech University, Water Resources Center, P.O. Box 4630, Lubbock, Texas 79409, USA.

Béchaux, J. Champ d'application et exemples de realisations de transports d'eau à grande distance (Field of application of the transport of water through great distances, with examples), *Houille Blanche,* **22,** 357–364 (1967).

Belyaev, I. P. *Interbasin transfer of water resources,* Working Paper 56. Original: English, UNDP/UN Interregional Seminar on River Basin and Interbasin Development, Budapest, Hungary, UN Dept of Economic and Social Affairs, (16–26 September 1975).

Cassidy, William F. *Water for cities,* Proc. 4th Am. Water Resources Conf., D. 34–41 (1968).

Clyde, George d. *The need for interstate water resource planning,* Proc. Western Interstate Water Conf., pp. 117–124 (1964).

Combes, G. and Saint-Laurent, F. Le transport de l'eau a grande distance (Water transport through great distances), *Houille Blanche,* **22,** 343–356, Summary: (In English) **22,** 356 (1967).

Cummings, Ronald G. *Interbasin Water Transfers—A Case Study in Mexico,* published by Resources for the Future Inc., distributed by Johns Hopkins Press, Baltimore and London, (1974).

Dersch, Eckhart *Water resources development—the river and the multi-basin region,* Cornell Univ. Water Resources Center Pbln. No. 11, pp. 1–15 (1965).

Domokos, Miklos *Evaluation of the effects of water transfer,* Working Paper 25, UNDP/UN Interregional Seminar on River Basin and Interbasin Development, Budapest, Hungary, UN Dept of Economic and Social Affairs (16–26 September, 1975).

Gardner, B. Delworth and Fullerton, Herbert H. Transfer Restrictions and the Misallocation of Irrigation Water, *Am. J. agric. Econ.* **50** (3), 556–71 (1968).

Hare, J. H. *Interbasin transport and water export,* Water Studies Inst. Colloq., Saskatoon, 24 pp. proc., (1971).

Hartman, L. S. and Seastone, Don, *Water Transfers: Economic Efficiency and Alternative Institutions,* Resources for the Future, Baltimore, (1970) 127 pp.

Horton, Robert E. Water division between drainage basins, *J. Am. Wat. Wks Ass.,* **24,** 1623–1683 (1932).

Howe, Charles W. Interbasin transfers: Do we need them? *Water Resour. Bull.* **7,** 670–675 (1971).

Howe, Charles W. and Easter, K. William, *Interbasin Transfers of Water,* Johns Hopkins Press, Baltimore, (1971) 196 pp.

Howson, Louis R. Factors affecting long-distance transmission of water, *J. Am. Wat. Wks Ass.,* **49,** 1359–1368 (1957).

Kuiper, Edward, Feasibility of water export, *Proc. Am. Soc. civ. Engrs,* **94** (HY-4), 873–891, DISC: **95** (HY-2), 727–728 (1968).

Lawson, J. D. Developments in water handling and management, *Aust. civ. Engng Constr.,* **10,** 29–34 (February 1969).

Laycock, Arleigh H. Canadian water for Texas? *Water Resour. Bull.,* **6,** 542–549 (1969).

Laycock, Arleigh H. *Interbasin transfer and Canadian water export?* Seminar, Johns Hopkins Univ., Baltimore, 25 pp. proc., Condensed reprint: *Reclamation* **12** (Spring), 3 (1971).

Maxwell, John C. Prospects for future water supply in the United States, *J. Am. Wat. Wks Ass.,* **57,** 549–553 (1965).

Mitchell, Grant C. *Interbasin transfer and export of water,* Ann. Conf. Canadian Water Resources Assn., Regina, 3 pp. proc., (1971).

Peterson, Dean F. *Philosophical consideration of water resource importation,* AAAS Symposium on Arid Lands, Dallas, Texas, pp. 398–405 processed, (30 December, 1968).

Poll, Charles, *Water Import Systems for Arid Land Development,* Int. Symposium on Increasing Food Production in Arid Lands, Monterrey, Mexico, pp. 43–54 (1969).

Quinn, Frank, Water transfers: Must the American west be won again? *Geogr. Rev.* **58,** 108–132 (1968).

Quinn, Frank *Images of Continental Water Engineering,* Seminar on the Continental Use of Arctic-Flowing Rivers, Washington State Univ., pp. 29–48 (1968).

Quinn, Frank J. *Area-of-origin protectionism in western waters,* Doctoral Dissertation, Univ. of Washington, 278 pp. +plates, (1970).

Quinn, Frank J. *Area-of-origin protectionism in western waters,* Environment Canada Social Science Series No. 6 (1973), 95 pp. +plates, Abst: *Reclamation* **13** (4), 11 (1974).

Reynolds, Robin R. *The water grid concept,* IIASA Task Force Meeting on Interregional Water Transfers, Laxenburg, (11–14 October, 1977).

Schramm, Gunter, Regional versus inter-regional efficiency in resource allocations, *Ann. Reg. Sci.,* **4,** 1–14 (December 1970).

Tinney, E. Roy, Criteria for analysis of interregional transfers of water, *J. Am. Wat. Wks Ass.,* **58,** 1369–1373 (1966).

Tinney, E. Roy, Water for western federal irrigation projects, in Campbell and Sylvester, *Water Resources Management and Public Policy,* Univ. of Washington Press, (1968) pp. 3–11.

Voropaev, Grigory V. *The scientific principles of large-scale areal redistribution of water resources in the USSR,* IIASA Task Force Meeting on Interregional Water Transfers, Laxenburg, (11–14 October, 1977).

Warnick, C. C. *Historical background and philosophical basis of regional water transfer*, AAAS Symposium on Arid Lands, Dallas, Texas, 30 pp., processed pp. 340–352 (30 December, 1968).

Whetstone, George A. Water power for the desert, *Water Power*, **13**, 388–390 (1961).

Whetstone, George A. *Interbasin diversion of water*, Proc. 4th Am. Water Resources Conf., pp. 490–498 (1968).

Whetstone, George A. Water as a commodity in international trade, *Water Resour. Bull.*, **5**, 65–71 (1969).

Whetstone, George A. Water diversion à la Française, *The Cross Section*, **15**, 2, 4 (15 April, 1969).

Whetstone, George A. *Continental water schemes*, Proc. ICASALS Arid Lands Symposium, Texas Tech, Lubbock, Texas, (1969).

Whetstone, George A. *The role of water exportation in national resources planning*, Engng Inst. of Canada, Annual Gen. Meeting, Ottawa, 12 pp. Proc., (1970).

Whetstone, George A. *Interbasin Diversion of Water: An Annotated Bibliography*, Vol. 1, WRC 70-2, Reproduced by the Texas Water Development Board, (1970).

Whetstone, George A. *Interbasin Diversion of Water: An Annotated Bibliography*, Vol. 2, WRC 71-1, Reproduced by the Texas Water Development Board, (1971).

Александровский, Ю. В., Баранов, В. А., Бостанджогло, А. А., Обрезков, С. С. и Шимельмиц, И. Я. Проблемы территориального перераспределения стока, *Труды IУ Всесоюзного Гидрологического Съезда*, Том **4**, Ленинград (1976).

Aleksandrovskiy, Yu. V., Baranov, V. A., Bostandzhoglo, A. A., Obrezkov, S. S. and Shimelmits, I. Ya. The problems of large-scale areal redistribution of runoff, *Tr. IV Vses. Gidrol. S''êzda*, Vol. **4**, Leningrad (1976).

Вендров, С. Л. Проблемы преобразования речных систем, *Гидрометеоиздат*, (1970).

Vendrov, S. L. The problem of river systems transformation, *Hydrometeoizdat*, (1970).

Вендров, С. Л. (Редактор) *Инженерно-географические проблемы проектирования и эксплуатации водохранилищ*, Наука, Москва, (1972).

Vendrov, S. L. (ed.) *Engineering and Geographical Problems in the Design and Operation of Water Reservoirs*, Nauka, Moskva (1972).

Воропаев, Г. В. Основные варианты территориального перераспределения водных ресурсов и связанные с ними научные проблемы, В сб.: *Оценка возможных изменений режима низовьев и устьев рек Арктической зоны Западной Сибири под влиянием водохозяйственных мероприятий*, Ленинград, (1976).

Voropaev, G. V. Principal alternatives of water resources reallocation and related scientific problems, *Otsenka Vozmozhnykh Izmeneniy Rezhima Nizovyev i Ustyev rek Arkticheskoy Zony Zapadnoy Sibiri pod Vliyaniem Vodokhozyaystvennykh Meropriyatiy*, Leningrad, (1976).

Воропаев, Г. В. Задачи и организация научных исследований в связи с проблемой перераспределения водных ресурсов, *Водные ресурсы*, № 3, (1976).

Voropaev, G. V. The objectives and organization of research in connection with the problem of water resources redistribution *Vodn. Resur*. 3, (1976).

Грин, Г. Б. Об основных принципах разработки научных основ пространственного управления водными ресурсами, *Бюллетень по водному хозяйству СЭВ*, 16, (1975).

Grin, G. B. The basic principles of developing the scientific fundamentals of areal water resources management, *Bull. po Vodnomu Khozyaystvu SEV*, 16, (1975).

Дунин-Барковский, Л. В., Воропаев, Г. В., Бостанджогло, А. А., Ауце, Д. А. и Грин, Г. Б. *Территориальное перераспределение стока; исследования, масштабы, альтернативы*, XXIII Международный географический конгресс, Симпозиум: "Преобразование водного баланса под влиянием хозяйственной деятельности", Ленинград, (1976).

Dunin-Barkovskiy, L. V., Voropaev, G. V., Bostandzhoglo, A. A., Autse, D. A. and Grin, G. B. *Large-scale areal streamflow redistribution: investigations, extent and alternatives*, XXIII Mezhdunarodnyi Geograficheskiy Kongress, Simposium, Preobrazovaniya Vodnogo Balansa pod Vliyaniem Khozyaystvennoy Deyatelnosti, Leningrad, (1976).

Кулаков, В. В. Соотношение между тепловыми и водными ресурсами Европейской территории СССР в связи с перераспределением речного стока, Сб. *Формирование ресурсов вод суши*, Москва, (1972).

Kulakov, V. V. The relationship between the thermal and water resources of the USSR European part in connection with river flow redistribution, *Formirovaniye Resursov vod Sushi*, Moskva, (1972).

Проблемы регионального перераспределения речного стока, *Труды Ленинградского Политехнического института*, **№ 351**, (1976).

The problems of regional river flow redistribution, *Tr. Leningr. Politekh. Inst*. No. **351**, (1976).

Украинский, Ф. Я. К вопросу о народохозяйственном обосновании крупных перебросок стока, *Труды Гидропроекта*, сб. 29, (1973).

Ukrainskiy, F. Ya. Economic grounds for large-scale streamflow diversion, *Tr. Gidroproekta*, issue 29, (1973).

Фиалковский, П. Г. О единой водохозяйственной системе, *Гидротехника и мелиорация*, № 1, (1976).

Fialkovskiy, P. G. An integrated water supply system, *Gidrotekh. Melior*. No. 1, (1976).

REGIONAL PROJECTS

Africa

Anon., Projet d'un canal dans la Sahara (A proposed canal in the Sahara), *Nature, Paris* **3** (11), 141–142 (1875).

Anon., Un projet pour la mise en valeur du Sahara: La mer conduite a 400/km à l'interieurdu desert (A project for developing the Sahara: The sea led 400km into the desert), *Usine Nouv.* **11,** 13, 15, 17, 19 (Juin 1955).

Ball, John, The Qattara Depression of the Libyan Desert and the possibility of its utilization for power-production, *Geogr. J.*, **82,** 289–314 +plates, (1933).

Bassler, Friedrich, Scheme for the Qattara Depression, *Water Power,* **20,** 494–498 (1968).

Bassler, Friedrich, New proposals to develop the Qattara Depression, *Water Power,* **27,** 238–243, 291–297 (1975).

Government of South Africa, Interbasin water transfers in South Africa, *Aqua,* **1** (1/2), 242 (1977).

Guisan, R. La mer interieure d'Algerie (Algeria's interior sea), *Bull. Soc. Vaudoise Ing. Archit.* (*Bull. Tech. Suisse Romande*), **10** (1–4), (1884).

Halwani, Salah, Janod, Albert, and Lajeat, J. P. Le percement du tunnel d'Awali de l'amenagement du Litani au Liban (The driving of the Awali tunnel of the development of the Litani in Lebanon), *Travaux,* **49,** 247–260 (1965).

Hauet, A., Note sur la mer interieure (Note on the interior sea), *Mem. C. r. Trav. Soc. Ing. civ. Fr.* **11,** 110–119 (1883).

Helene, Maxime, La mer interieure d'Algerie (Algeria's interior sea), *Nature, Paris* **11,** 22–24, 35–38 (1882).

De Martino, Guiseppe, The Qattara Depression, *Water Power,* **25,** 27–31 (1973).

Matthews, P. E. Diversion of water to the upper region of the Tugela Basin, *Natal Town Reg. Rep.*, **16,** 30 pp. (1969).

Midgley, D. C. Apprisal and management of water resources in South Africa, *S. Afr. J. Sci.*, **67,** 210–217 (1971).

Myburgh, R. I. D. McC., The water projects of the Namib Desert, *S. Afr. J. Sci.* **67,** 152–158 (1971).

Van Robbroeck, T. P. C. *Inter basin water transfers in South Africa,* United Nations Water Conf., Mar del Plata, Argentina, Conf. Document E/CONF. 70 TP 223, 20 pp. (1976).

Roudaire, E. Une mer interieure en Algeria (An interior sea in Algeria), *Revue Deux Mondes,* 323–350 (15 Mai 1874).

Roudaire, E. La mer interieure Africaine (The African interior sea), *Mem. C. r. Trav. Soc. Ing. civ. Fr.*, **11,** 484–515, Disc: 482–483, 595–624 (1883).

Sirry, H. The Qattara power scheme, *Geogr. Rev.*, **19,** 290–292 (1929).

Soergel, Herman *Lowering the Mediterranean: irrigating the Sahara* (Panropa Project), Gebhardt, Leipzig, (1929) 48 pp.

Stallebrass, J. L. Regional water resources planning in South Africa, *S. Afr. J. Sci.*, **67,** 159–165 (1971).

Thorrington-Smith, E. *Toward a Plan for the Tugela Basin,* Town and Regional Planning Commission, Natal, Pietermaritzburg, (1960) 338 pp.

Wellington, John H. Zambesi–Okovango development projects, *Geogr. Rev.*, **39,** 552–567 (1949).

Whetstone, George A. *The Tugela–Vaal: a precedent-shattering diversion,* Short Papers: 8th Am. Water Resources Conf., pp. 14–15 (1972).

Whetstone, George A. Adaptation to aridity: water development in South Africa, *Water Resour. Bull.*, **8,** 600–607 (1972).

North America

Adams, Thomas C. Development of the Colorado River in the Upper Basin, *Trans. Am. Soc. civ. Engrs,* **105,** 1345–1365, Disc: **105,** 1366–1375 (1940).

Anon., The lake in the Colorado Desert, *Engng News, N.Y.*, **26,** 313–314 (1891).

Anon., The development of water power on the Chicago drainage canal, *Engng News, N.Y.*, **38,** 152 (1897).

Anon., The diversion of interstate waters, *Engng News, N.Y.*, **46,** 369–370; **47,** 301, 310 (1901).

Anon., The five dams and wood stave conduit of the Southern California Mountain Water Co., *Engng News, N.Y.*, **51,** 335–337 (1904).

Anon., California's long-range water plan, *Engng News Rec.*, **119,** 344–348 (1937).

Anon., Southwest plan would create regional water network, *Engng News Rec.*, **171,** 18–19 (5 September 1963).

Anon., Southwest should tap the Snake River, *Engng News Rec.*, **171,** 56–57 (14 November 1963).

Anon., Udall mediates Pacific Southwest water plan, *Engng News Rec.*, **172,** 14–15 (9 January 1964).

Anon., Water for all: Col. F. Z. Pirkey's plan (Western Water Project), *West. Wat. News,* p. 2 (16 January 1964).

Anon., Fryingpan–Arkansas Project, *West. Constr. News,* **42,** 45–52 (July 1967).

Anon., *South Saskatchewan River Project,* Queen's Printer, Cat. No. A22-7368, Ottawa, 32 pp., litho, (1967).

Anon., Six-state water plan proposed, *Am. Cy Meg.,* **83,** 164 (June 1968).

Anon., Water diversion study fires uproar in Ontario, *Engng News Rec.,* **183,** 23 (24 July 1969).

Armstrong, Ellis L. The undersea aqueduct: A new concept in transportation, *Proc. Am. Soc. civ. Engrs,* **98** (TE 2), 303–310 (1972).

Banks, Harvey O. *et al., The California Water Plan,* California Dept of Water Resources, Bull. No. 3, 246 pp. +plates, (1957).

Basic Facts Booklet, *California Water Project,* California Dept of Water Resources, Division of Operations and Maintenance, (April 1973).

Bayley, Fred H. III, Possible diversion of Mississippi River flows, *Proc. West Texas Wat. Conf.,* **9,** 19–25 (1971).

Bayley, Fred. III, *West Texas and Eastern New Mexico water import study,* Paper, Am. Water Resources Conf., Baton Rouge, 34 pp. (1975).

Beck, R. W. and Associates, *A new water plan for the Great Plains,* Paper presented at Mid-West Elect. Consumers Assn., 50 pp., Proc., (8 December 1967).

Boswell, Howard B. *The Texas Water Plan,* pp. 24–29 (1969).

Branch, L. V. The Colorado River Aqueduct: Aqueduct operating program, *Engng News Rec.,* **121,** 677–679 (1938).

Burdick, Quentin N. *Garrison diversion unit, Missouri River Basin project,* Senate Report No. 470, 89th Congr., 1st Session, 15 pp. (1965).

Burleigh, Harry P. *The bureau of reclamation and the Mississippi River export plan.* Symposium on Aspects of the Diversion of Mississippi River Water to Texas and New Mexico, pp. 95–106 (1969).

Burleigh, Harry P. Export of Mississippi River water to Texas and New Mexico, (a portion of a Symposium), *J. Am. Wat. Wks Ass.,* **62,** 367–369 (1970).

Campbell, Donald E., Marlow, Thomas A. and Wright, Roy L. Homestake delivery system pipeline, *Proc. Am. Soc. civ. Engrs,* **93** (PL-2), 71–83, Disc: **94** (PL-1), 187–189 (1967).

Cannon, D. E. *San Juan–Chama Project,* Proc. 14th New Mexico Water Conf., pp. 55–59 (1969).

Cassidy, William F. New England water supply, *J. Am. Wat. Wks Ass.,* **60,** 21–24 (1968).

Clark, Robert D. *The embattled colossus: New York City water supply system, its past, present and future,* Proc. 4th Am. Water Resources Conf., pp. 25–33 (1968).

Clinton, F. M., Director, Region 6, USBR, *Garrison Diversion Unit,* 86th Congr., 2nd Session House Doc. 325, 229 pp. +plates, (1960).

Cole, Donald B. Transmountain water diversion in Colorado, *Colo. Mag.,* **25,** 49–65, 118–135 (1948).

Cory, H. T. Closing the break of the Colorado River into the Salton Sink, Southern California, *Engng News, N.Y.,* **56,** 671–674, Edit: **56,** p. 687 (1906).

Criddle, Wayne D. *Water resources and needs in areas of deficiency—Utah,* Proc. Western Interstate Water Conf., pp. 50–57 (1964).

Cummings, Ronald G. *Interbasin Water Transfers: A Case Study in Mexico,* published by Resources for the Future Inc., distributed by Johns Hopkins Press, Baltimore and London, (1974).

Dealey, J. Q., Jr, The Chicago Drainage Canal and St Lawrence Development, *Am. J. Int. Law,* **23,** 307–328 (1929).

Debler, E. B. Inter-water-shed diversions by tunnels on United States reclamation projects, *Trans. Am. Soc. civ. Engrs,* **94,** 346–351, Disc: 368 (1930).

Dominy, Floyd E. *Water supplies for the Southwest—What of the future?* Proc. 13th Annual Water Conf., Las Cruces, NM, pp. 31–37 (28–29 March 1968).

Dunn, William G. *The modified Snake-Colorado project,* Hearings before the Subcommittee on Water and Power Resources of the Committee on Interior and Insular Affairs, US Senate, 90th Congr., 1st Session, on S.1004, S.1013, S.861, S.1242 and S.1409, pp. 718–728 (1967).

Eckart, Nelson A. The water supply system of San Francisco, *J. Am. Wat. Wks Ass.,* **32,** 751–794 (1940).

Edmonston, A. D. California proposes diversion of Feather River water to Southern California, *Civ. Engng,* **23,** 229–231 (1953).

Eggleston, Wilfrid, The South Saskatchewan River Project, *Can. geogr. J.,* **46,** 232–243 (1953).

Elder, C. C. The Colorado River Aqueduct as a supplement to local water-sources, *Trans. Am. geophys. Un.,* **23,** 23–103 (1942).

Engelbert, Ernest A. The origins of the Pacific Southwest water plan, *West. Resour. Pap.* **6,** 125–256 (1964).

Follansbee, Robert, Upper Colorado River and its utilization, *U.S. Geol. Surv. Water Supply Pap.,* **617,** 394 +plates, (1929).

Follansbee, Robert, Trans-mountain diversions in Colorado, *Trans. Am. Soc. civ. Engrs,* **94,** 359–367 (1930).

Freeman, John R. *et al.* Report on the proposed 226-mile aqueduct for the water supply of Los Angeles, Cal., *Engng News, N.Y.,* **57,** 93–97, Edit: **57,** 109, Letter to Ed., Wm. Mulholland, p. 329 (1907).

Freese, Nichols and Endress, *Report to Texas Water Development Board on Interbasin Canals and Pipelines,* 192 pp. +maps, (1967).
Garduno, Hector, Mestre, Eduardo and Tapia, Francisco, *Large-scale transfers within Master water planning in Mexico,* IIASA Task Force Meeting on Interregional Water Transfers, Laxenburg, (11–14 October 1977).
Gill, G. S., Gray, E. L. and Secklery, D. The California Water Plan and its critics: A brief review, in Seckler, D. (ed.) *California Water, A Study in Resource Management,* Univ. of California Press, Berkeley, CA, (1971).
Heinly, Burt A. The completion of the Los Angeles Aqueduct, *Engng News,* **69,** 1257–1268; **70,** 920–923; Edit: **69,** 1289, Disc: **70,** 318–320, 1095 (1913).
Hinds, Julian, Colorado river water for California, *Civ. Engng,* **7,** 573–575 (1937).
Hyatt, Edward, The Central Valley Project of California, *Civ. Engng,* **4,** 482–486, Disc: **5,** 31, 32, 180 (1934).
Hyatt, Edward, The Central Valley Project of California, *J. Am. Wat. Wks Ass.,* **30,** 389–401 (1938).
Iorns, W. V., Hembree, C. H. and Oakland, G. L. Water resources of the Upper Colorado River Basin—technical report, *U.S. Geol. Surv. Prof. Pap.,* **441,** 370 pp. +plates, (1965).
Johnson, Ralph B. *The area of origin and a Columbia River diversion,* State of Washington, Water Res. Center Rep. No. 5B, 41 pp. (1971).
Kelly, Roland P. *North American Water and Power Alliance—NAWAPA,* Proc. 3rd Am. Water Resources Conf., pp. 28–35 (1967).
Kendrick, John J. *Water importation to West Texas and New Mexico,* Proc. 13th Annual Water Conf., Las Cruces, NM, pp. 18–24, (28–29 March 1968).
Kierans, T. W. *The Great Replenishment and Northern Development Canal,* Joint Meeting EIC–ASCE–ICE, Toronto, 11 pp. Proc., (1965).
Kierans, T. W. *The Grand Canal concept: A means of linking the fresh water needs of the Great Lakes and the continent to the development of Northern Canada,* Symposium on The Fresh Water of New York State: Its Conservation and Use, State University of New York, Buffalo, pp. 87–98 (13–17 June 1966).
Knight, Oliver, Correcting nature's error: The Colorado–Big Thompson project, *Agric. Hist.,* **30,** 157–169 (1956).
Kuiper, E. Canadian water export, *Engng J.,* **49,** 13–18, Disc: 32–34 (1966).
Kuiper, E. Water utilization in Canada, *Engng Dig., Toronto,* **13,** 66–69 (13 June 1967).
Laycock, Arleigh H. *The rationale of water export and its implications for water resources in the north,* Proc. 2nd Northern Resources Research Conf., Whitehorse, Yukon, pp. 90–96 (1968).
Laycock, A. H. Water, Chapter 5 in Warkentin, *Canada, a Geographical Interpretation,* Methuen, Toronto, (1968) pp. 112–136.
Laycock, A. H. Canadian water for Texas? *Water Resour. Bull.,* **6,** 542–549 (1969).
Laycock, A. H. *Water development in Alberta's future: Problems and prospects—the demand for water export in the year 2000 A.D.,* Proc. Univ. of Calgary Seminar on The Environment of Alberta—Past, Present and Future, 7 pp. proc., (1971), Condensed reprint: *Reclamation,* **11** (2), 6 (1971).
Lee, F. C. and Stern, D. *A feasibility study of mainland shelf undersea aqueduct,* Report to the US Dept of the Interior, National Engineering Science Co., Pasedena, CA, (1965).
Lippincott, J. B. Southwestern border water problems, *J. Am. Wat. Wks Ass.,* **31,** 1–28 (1939).
Luther, Lawrence A. Reaching 240 miles across a desert for water, *Compress. Air Mag.,* **39,** 4452–4458 (1934).
McCammon, L. B. and Lee, F. C. Undersea aqueduct system, *J. Am. Wat. Wks Ass.,* **58** (7), 885–892 (1966).
McCasland, S. P. Water from Pacific Northwest for deserts of Southwest, *Civ. Engng,* **22,** 143–146 (1952).
McFarland, K. D. and Kelly, Roland P. *The NAWAPA concept,* Proc. 3rd West Texas Water Conf., pp. 125–133 (1965).
Magnusson, Knut, *Should Water in Western Canada be Exported to U.S.A.?* Mimeo., 31 pp. +plates (2nd edn.), (13 January 1966).
Merriman, R. M. Driving 91 miles of tunnels on the Colorado River aqueduct, *Engng News Rec.* **113,** 97–105 (1934).
Moss, Frank E. *et al. Western water development: A summary of water resources projects, plans and studies relating to the Western and Midwestern United States,* Committee Print, 88th Congr., 2nd Session, 101 pp., (1964).
Moss, Frank E. *Future U.S. water resources development,* Symposium (E. Roy Tinney, ed.) The Continental Use of Arctic-Flowing Rivers, Washington State Univ., pp. 133–151 (1968).
Nelson, Samuel B. *Snake-Colorado Project: A plan to transport surplus Columbia River Basin water to the arid Pacific Southwest,* Dept of Water and Power, City of Los Angeles, (1963).
Nelson, Samuel B. Increase water supply—don't allocate shortage, *Am. Cy. Mag.* **80,** 96–98 (March 1965).
Nelson, Samuel B. *Snake Colorado project,* Am. Soc. civ. Engrs Nat. Meeting, San Diego, CA, 16 pp. processed, (1968).
Van Norman, H. A. The proposed Los Angeles–Colorado River aqueduct, *Trans. Am. Soc. civ. Engrs,* **94,** 352–358, Disc: 368–376 (1930).

Northern Ontario Water Resources Studies, *Summary report on engineering feasibility and cost investigations,* Environment Canada, Inland Waters Directorate, Water Resources Branch, Ottawa, Canada, (1973).

Panero, Robert B. and Candella, Basil J. *Super projects,* Proc. 4th Am. Water Resources Conf., pp. 208–217 (1968).

Parker, Garald G. *et al.* Water resources of the Delaware River Basin, *U.S. Geol. Surv. Prof. Pap.,* **381,** 200 pp. +plates, (1964).

Parsons, The Ralph M. Company, NAWAPA: North American Water and Power Alliance, *North American Water and Power Alliance Conceptual Study,* Vol. 1, engineering. Ref. No. 606-2934-001, Los Angeles, (1964).

Parsons, The Ralph M. Company, *NAWAPA: North American Water and Power Alliance,* Parsons' Brochure 606-2934-19, Los Angeles, 50 pp., (1967).

Parson, The Ralph M. Company, *North American Water and Power Alliance Conceptual Study,* Vol. 1, engineering, (1974).

Parsons, The Ralph M. Company, *North American Water and Power Alliance Summary Report for Alaska,* (January 1977).

Preston, Porter J. The Colorado–Big Thompson Project, *Colo. civ. Engng,* **8,** 517–519 (1938).

Price, Bobby E. (ed.), *Aspects of the diversion of Mississippi River water to Texas and New Mexico,* Symposium, Louisiana Polytechnic Inst., 160 pp., (19 June 1969).

Price, Bobby E. Possible diversion of Mississippi River water to Texas and New Mexico, *Water Resour. Bull.,* **7,** 676–683 (1971).

Primus, C. L., Paul, L. and Anderson, I. H. *Water diversion proposals of North America,* Alberta Dept of Agric., Water Resources Div. 42 pp. +maps, (1969).

La Rue, E. C. Colorado River and its utilization, *U.S. Geol. Surv. Water Supply Pap.,* **395,** 231 pp. +plates, (1916).

Schneider, William J. Water data for metropolitan areas, *U.S. Geol. Surv. Water Supply Pap.* **1871,** 397 pp., (1968).

Sewell, W. R. Derrick, NAWAPA: A continental water scheme (symposium); pipedream or practical possibility? *Bull. atom. Scient.* **23,** 9–13 (23 September 1967).

Smith, Joe H. Jr, Water for a thirst continent: A condensation of the "Western States Water Augmentation Concept" by Lewis G. Smith, *Irrig. Age,* **3,** 10–15 (3 April 1969).

Smith, Lewis G. *Western States Water Augmentation Concept,* Fedn. of Rocky Mtn. States, Denver, 44 pp. +plates, (1967).

Smith, Lewis G. Toward a national water plan, Supplement to April issue of *Irrig. Age.* 24 pp. (1969).

Smith, Lewis G. *Concept for a Pacific Aqueduct System,* Denver, Colorado 8 pp. +plates, proc., (1972).

Smith, Stephen C. and Brewer, Michael F. *California's man-made rivers: History and plans for water transfers,* Div. of Agric. Sciences, Univ. of California, Berkeley, (1961).

Stam, G. G. *Regional development of the waters of the Colorado River through interbasin diversions,* UN Dept of Econ. and Soc. Affairs, International Seminar on River Basin and Interbasin Development, Budapest, Hungary, (16–26 September 1975).

Stetson, T. M. *Review of Pacific Southwest Water Plan,* Consulting Engineer report to Six-Agency Committee, Los Angeles, CA, (1964).

Symposium, Water needs of the Pacific slope, *J. Am. Wat. Wks Ass.,* **57,** 1238–1272 (1965).

Texas Board of Water Engineers and Three Federal Agencies, *Water developments and potentialities of the State of Texas,* Sen. Doc. No. 111, 85th Congr., 2nd Session, 175 pp. +plates, (1958).

Texas Water Development Board, *Water for Texas: A plan for the future (Preliminary),* TWDB, viii+32 pp. +plates, (1966).

Texas Water Development Board, *The Texas Water Plan: Summary,* TWDB, Austin, xiii +50pp. +plates, (November 1968).

Texas Water Development Board, *The Texas Water Plan,* TWDB, 207 pp. + plates, (1968).

Thon, J. George and Campbell, Donald E. Homestake trans-divide water supply project, *Proc. Am. Soc. civ. Engrs,* **93** (PL-3), 1–14; Disc: **94** (PL-1), 195–196; **95** (TE-2), 389–390 (1967).

Tinney, E. Roy, *A broad view of the problem,* Symposium on Northwest–Southwest Water Diversion Issues, Oregon States Univ., pp. 1–10 (1967).

Tinney, E. Roy (ed.), Symposium on The Continental Use of Arctic-Flowing Rivers, Washington State Univ., 176 pp. (1968).

Tinney, E. Roy and Quinn, Frank J. *Canadian waters and arid lands,* AAAS Symposium on Arid Lands, Dallas, Texas, pp. 411–715, processed, (31 December 1968).

Trommershausen, W. E. *A new water resource plan for the Great Plains including importation from Canada* Western Water Resources Inst., Santa Fe, NM, 21 pp. + plates, (1969).

US Bureau of Reclamation, Region 6, *Supplemental report on Garrison diversion unit (initial stage—250,000 acres),* USBR, Bismarck, ND, 61 pp. +maps, (revised February 1965).

US Bureau of Reclamation, *California undersea aqueduct pre-reconnaissance study,* US Dept of the Interior, Bureau of Reclamation, 106 pp. (1969).

US Bureau of Reclamation, *West Texas and Eastern New Mexico import project,* USBR, 202 pp. +plates, (1973).

US Department of the Interior, *Pacific Southwest Water Plan: report and appendix,* (August 1963).

Vandertulip, John J. *A plan for meeting the 1980 water requirements of Texas,* Texas Board Wat. Engrs., 198 pp. +plates, (1961).

Vandertulip, John J. *Water importation—current outlook,* Proc. 6th West Texas Water Conf., pp. 22–32 (1968).

Vandertulip, John J. and Seward, Lewis B. *The preliminary Texas water plan,* Texas Water Dept Board Rep. 31, pp. 1–28 (1966).

Watts, Calvin T. and Lafleur, Robert A. *Proposed diversion of Mississippi River water from the viewpoint of the basin of origin,* Symposium on Aspects of the Diversion of Mississippi River Water to Texas and New Mexico, Ruston, LA, pp. 9–23 (19 June, 1969).

Wells, Dan M. *Irrigation in the Texas High Plains and the Texas water plan,* Am. Soc. civ. Engrs, Irrig. and Drain. Conf., (1969) 16 pp. Proc.

West Texas and Eastern New Mexico, *Import project executive summary,* US Dept of the Interior, Bureau of Reclamation, (June 1973).

Weymouth, Frank E. The proposed Colorado River aqueduct and metropolitan water district, Abst: *Proc. Am. Soc. civ. Engrs,* **56,** 1283–1291 (1930).

Weymouth, F. E. The Colorado River aqueduct: Urgent need for the aqueduct, *Engng News Rec.,* **121,** 639–645 (1938).

Williams, R. B. The all-American canal is materializing, *Reclam. Era,* **25,** 162–164 (1935).

South America

Ackerman, Adolph J., Matthias, Franklin T. and Vogan, George O. Brazil's Paraiba–Pirai diversion project includes world's largest underground power station, *Civ. Engng,* **21,** 652–656, Correction: **22,** 112–113 (1951).

Anon., Irrigation projects in Peru, Based on reports of Charles W. Sutton, *Georgr. Rev.* **17,** 487–490 (1927).

Anon., The Paraiba–Pirai Diversion, *Water Power,* **5,** 286–293 (1953).

Anon., Choco development project, *Water Power,* **19,** 6, Editorial notes: 2, 100 (1967).

Cote, E. F. Quelques projets gigantesques (Some gigantic projects), *Houille Blanche* (Old Series), **5,** 65–67 (1906).

Enock, C. Reginald, Water power in the Andes, *The Engineer,* **106,** 313 (1908).

Glassner, Martin I. The Rio Lauca: Dispute over an international river, *Geogr. Rev.* **60,** 192–207 (1970).

Green, D. W. Chira-Piura: A project to improve irrigation in Northern Peru, *Water Power,* **27,** 309–318 (1975).

Mary, M., Le Perou: Ses resources hydroelectriques (The hydroelectric resourses of Peru), *Houille Blanche,* **14,** 450–456 (1959).

Meischeider, Herbert, The Jequetepeque multipurpose scheme in Peru, *Water Power,* **26,** 178–179 (1974).

Ministerio de Obras Publicos y Transportes, *El sistema Rio de la Unidad, esquema alternativa de analisis no. 101,* Oficinas Proyecto Rio de la Unidad, Direccion General de Aguas, Santiago, Chile, (1972).

Panero, Robert, *A South American "Great Lakes" system,* Report HI-788/3-RR, Hudson Inst. Inc., Croton-on-Hudson, NY, 29 pp. (1967).

Pieretti, J. G. and Cole, E. C. Caracas water supply lifted 3,125 ft over divide, *Civ. Engng,* **26,** 810–813 (1956).

Ribi, R. and Schlittler, F. Expanding the power supply for Lima, Peru, *Water Power,* **16,** 160–164, 243–250 (1964).

Riva, Jorge, J. C. Notas sobre un racional aprovechamiento de los recursos de agua superficial del extremo del continente Americano, entre los paralelos 38° y 49° S (Notes on a rational utilization of the surface water resources of the southern tip of South America, between the 38th and 49th parallels), *Water for Peace,* **8,** 594–608, English Summary, 607–608 (1967).

Tomasek, Robert D. The Chilean-Bolivian Lauca River dispute and the O.A.S., *J. Inter-Am. Stud.* **9,** 351–366 (1967).

Asia

Anon., The Indus Basin Project, Pakistan. An engineering solution to a political problem, *Wat. and Wat. Engng,* **71,** 441–452 (1967).

Buck, A. C. The Johore River Project—Singapore, *J. Instn Wat. Engrs*, **23,** 13–57 +plates, (1969).

Hodges, R. C. Indus Water for Pakistan, *Can. geogr. J.*, **61,** 96–105 (1960).

Savage, Christopher D. W. and Xarpenter, Thomas G. Indus Basin settlement scheme: T-S-M-B link canal system, *Proc. Instn. civ. Engrs*, **32,** 549–571, Disc: **35,** 184–204 (1965).

Syed, N. C. *Inter basin transfer of water resource: Case study of Indus project,* Working Paper No. 28, UNDP/UN Interregional Seminar on River Basin and Interbasin Development, Budapest, Hungary, United Nations Dept of Economic and Social Affairs, (16–26 September 1975).

Wiener, A., Irrigation water system—Israel's national water grid, *Water for Peace,* **7,** 195–205 (1967).

Australia

Anon., The Coolgardie Goldfields Water Supply, *The Engineer,* **85,** 57–58, 75–77 (1898).

Anon., The Coolgardie pipe line: A remarkable engineering problem, *Engng News, N.Y.*, **40,** 233–234 (1898).

Anon., The Snowy Mountains Scheme, *Engineering, Lond.* **182,** 232–235, 394–396, 630–633, 661–663 (1956).

Bradfield, J. J. C. *Watering Inland Australia,* Rydge's, pp. 586–589, 606 (1 October 1941).

Bradfield, J. J. C. Restoring Australia's parched lands, *Aust. Q.*, **14,** 27–39, (March), Disc: Leeper, G. W., **14,** 50–52, (June 1942).

Dann, H. E. *et al.*, The investigation, design, construction and commissioning of the Guthega Project of the Snowy Mountains scheme, *J. Instn Engrs Aust.*, **29,** 129–158 (1957).

Davies, I. A comprehensive water supply scheme—Western Australia, *J. Instn Engrs Aust.*, **21,** 1–9 (1949).

Hudson, Sir William, The Snowy Mountains hydro-electric project, *J. Instn Engrs Aust.*, **34,** 107–113 (1862).

Hunter, J. R. and Hartwig, W. P. The design and construction of the Tooma–Tumut project of the Snowy Mountains scheme, *J. Instn Engrs Aust.* **34,**, 163–185 (1962).

Idriess, Ion L. *The Great Boomerang,* Angus & Robertson, Sydney and London, (1945) 253 pp.

Loder, L. F. The Snowy River Scheme, *J. Instn Engrs Aust.*, **21,** 40–41 (1949).

Main, J. M. The Development of the Snowy River, *J. Instn Engrs Aust.*, **17,** 209–216 (1945).

Warrell, E. G. The Snowy Mountains Scheme—Australia, *Water for Peace,* **8,** 248–258 (1967).

Europe

Anon., Immense new water supply for Paris authorized, *Engng News Rec.* **108,** 316 (1932).

Van Hulten, M., Lelystad, Prospects for the development of inland water transport in Europe, *Geogr. J.*, **1,** (1977).

Rees, Judith, Thirsty land with a temperate climate, *Geogr. Mag.*, **42,** 136–145 (1969).

Schmidt, Fritz, Projection and development of long distance water supplies illustrated by the example of the Lake of Constance scheme, *Water for Peace,* **6,** 1003–1009 (1967).

Vie, Georges, Le Canal de Provence: Ses objectifs et sa construction (The Canal de Provence: Its objectives and its construction), *Genié civ.*, **144,** 545–548 (1967).

USSR

Anon. (C. M.), Des cours d'eau Russes seront detournes de l'Artique vers la Caspienne (Some Russian rivers are to be diverted from the Arctic to the Caspian), *Sci. Prog., Lond.*, **3325,** 221–222 (1962).

Anon. (C. M.), Pour maintenir le niveau de la Mer Caspienne (To maintain the level of the Caspian Sea), *Sci. Prog., Lond.*, **3343,** 453–454 (1963).

Anon. Objections contre l'amenagement de l'Ob (Objections to developing the Ob River), *Sci. Prog., Lond.*, **3344,** 492 (1963).

Anon. Du Ladoga à la Caspienne (From Lake Ladoga to the Caspian Sea), *Sci Prog., Lond.*, **3362,** 230 (1965).

Anon. Grandes derivations hydrauliques en projet en URSS (Large projected water diversions in Russia), *Construction,* **25,** 410 (1970).

Anon., Irrigation figures heavily in Russia's massive diversion program, *Wld Irrig.*, **18,** 14–16 (January 1968).

Dunin-Barkovsky, L. V., Voropaev, G. V. and Gerardi, I. A. The scientific basis for the territorial redistribution of the water resources of the USSR, Abst: *Aqua,* **1** (1/2), 207, (1977).

Micklin, Philip P. Soviet plans to reverse the flow of rivers: The Kama–Vychegda–Pechora project, *Can. geogr.*, **13,** 199–215 (1969).

Voropaev, Grigory V. *The scientific principles of large-scale areal redistribution of water resources in the USSR,* IIASA Task Force Meeting on Interregional Water Transfers, Laxenburg, (11–14 October 1977).

Аверьянова, А. Г. Бассейн Волги и его региональное водохозяйственное значение, *Водные ресурсы,* № 6, (1974).

Averyanova, A. G. The Volga Basin and its signficance as regional water supply, *Vodn. Resur.* No. 6, (1974).

Ануфриев, А. Ф., Братцев, А. П. Природные и экономические предпосылки переброски части стока р. Оби в бассейн Волги, Сб. *Проблемы энергетики Крайнего Севера,* ч. III, Якутск, (1975).

Anufriyev, A. F. and Brattsev, A. P. Environmental and economic circumstances of the diversion of part of the Ob River flow into the Volga Basin, *Problemy Energetiki Kraynego Severa,* Part III, Yakutsk, (1975).

Биндеман, Н. Н. К вопросу о перспективах использования подземных вод Казахстана в связи с проблемой переброски стока Сибирских рек, *Водныеи ресурсы,* № 4, (1975).

Bindeman, N. N. On the prospects of the utilization of Kazakhstan's groundwater in connection with the problem of diversion of the flow of the rivers of Siberia, *Vodn. Resur.* No. 4, (1975).

Бостанджогло, А. А. Проблемы перераспределения стока в СССР, *Труды координационного совещания по гидротехнике,* вып. **108,** (1976).

Bostandzhoglo, A. A. The problems of stream flow redistribution in the USSR, *Tr. Koord. Soveshch. po Gidrotekh.,* **108,** (1976).

Братцев, А. П. Новый взгляд на проблему переброски части стока р. Печоры на Юг, Сб. *Проблемы рационального использования естественных ресурсов и охраны природы Коми АССР,* Сыктывкар, (1975).

Brattsev, A. P. A new view to the problem of partial diversion of Pechora River runoff to the south, *Problemy Ratsionalnogo Ispolzovaniya Estestvennykh Resursov i Okhrany Prirody Komi ASSR,* Syktyvkar, (1975).

Вендров, С. Л. Некоторые аспекты межбассейнового перераспределения речного стока в Европейской части Союза и в Срединном регионе, Сб. *Влияние перераспределения стока рек бассейна Оби на природу тайги Срединного региона,* Иркутск, (1975).

Vendrov, S. L. Some aspects of the inter-basin redistribution of river flow in the European part and the Central region of the USSR,—*Vliyaniye Pererasspredeleniya Stoka Rek Basseyna Obi na Prirodu Taygi Sredinnogo Regiona,* Irkutsk, (1975).

Вендров, С. Л. Роль и интересы Западной Сибирской равнины в решении проблемы перебросок стока, Сб. *Вопросы антропогенных изменений водных ресурсов,* Москва, (1976).

Vendrov, S. L. The role and interests of the West Siberian Plain in the solution of the streamflow redistribution problem, *Voprosy Antropogennykh Izmeneniy Vodnykh Resursov,* Moskva, (1976).

Гангардт, Г. Г. К вопросу о переброске части неиспользуемого стока Северных и Сибирских рек в районы, испытывающие дефицит водных ресурсов, *Гидротехническое строительство,* № 8, (1971).

Gangardt, G. G. Diversion of part of the unused flow of the northern and Siberian rivers to areas with a shortage in water resources, *Gidrotekh. Stroit.,* No. 8, (1971).

Герарди, И. А. Проблема использования вод Обского бассейна для обводнения Казахстана и Средней Азии в комплексе с осушением Западно-Сибирской низменности, Сб. *Комплексное освоение водных ресурсов Обского бассейна,* Новосибирск, (1970).

Gerardi, I. A. The problem of utilization of the waters of the Ob Basin for irrigation in Kazakhstan and Central Asia, combined with the drainage of the West-Siberian Lowland, *Kompleksnoye Osvoeniye Vodnykh Resursov Obskogo Basseyna,* Novosibirsk, (1970).

Герарди, И. А. Воду сибирских рек засушливым землям юга, *Гидротехника и мелиорация,* № 12, (1972).

Gerardi, I. A. The rivers of Siberia must supply water to the arid areas of the south, *Gidrotekh. Melior.,* No. 12, (1972).

Герарди, И. А. О повышении водообеспеченности р. Волги и переброска части стока северных рек на юг, *Гидротехника и мелиорация,* № 11, (1973).

Gerardi, I. A. On the increase in water resources of the Volga and the partial diversion of the flow of the northern rivers to the south, *Gidrotekh. Melior.* No. 11, (1973).

Герарди, И. А. Природно-хозяйственные основы переброски части стока Сибирских рек в республики Средней Азии и маловодные области Казахской ССР, *Мелиорация почв Средней Азии, Казахстана и Западной Сибири в связи с переброской части стока Сибирских рек в южные районы страны,* Пущино, (1973).

Gerardi, I. A. Environmental and economic principles of the partial diversion of the flow of the Siberian rivers to the Central Asia Republics and the arid areas of the Kazakh SSR, *Melioratsiya pochv Sredney Azii, Kazakhstana i Zapadnoy Sibiri v Svyazi s Perebroskoy Chasti Stoka Sibirskikh rek v Yuzhnye Rayony Strany,* Pushchino, (1973).

Герарди, И. А. Единая государственная система регулирования и межбассейнового перераспределения водных ресурсов СССР, *Труды 1У Всесоюзного Гидродогического Съезда,* Том **4,** Ленинград (1976).

Gerardi, I. A. The state integrated system of water resources regulation and inter-basin redistribution in the USSR, *Tr. IV Vses. Gidrol. S"ëzda,* **4,** Leningrad (1976).

Герарди, И. А. Технико-экономические соображения по переброске части стока Сибирских рек в республики Средней Азии и маловодные районы Казахстана, Сб. *Оценка Возможных изменений режима низовьев и устьев рек Арктической зоны Западной Сибири под влиянием водохозяйственных мероприятий,* Ленинград, (1976).

Gerardi, I. A. Technical and economic considerations in diverting part of the flow of the Siberian rivers to the Republics of Central Asia and the arid areas of Kazakhstan, *Otsenka Vozmozhnykh Izmeneniy Rezhima Nizovyev i Ustyev rek Arkticheskoy Zony Zapadnoy Sibiri pod Vliyanien Vodokhozaystvennykh Meropriyatiy,* Leningrad, (1976).

Дунин-Барковский, Л. В., Воропаев, Г. В. и Герарди, И. А. *Научные основы территориального перераспределения водных ресурсов в СССР,* Тезисы доклада, TP 165, Конференция Организации Объединенных Наций по мировым водным ресурсам, Мар-дель-Плата, Аргентина, (14-25 марта 1977).

Dunin-Barkovsky, L. V., Voropaev, G. V. and Gerardi, I. A. The scientific basis for the territorial redistribution of the water resources of the USSR, United Nations Water Conference, Mar Del Plata, Argentina, 14–25 March, Abst: *Aqua,* **1** (1/2), 207 (1977).

Зарубаев, Н. В. Комплексные исследования по обоснованию проекта межбассейнового перераспределения части стока Сибирских рек, *Труды Ленинградского политехнического института,* № **346,** (1976).

Zarubaev, N. V. Interdisciplinary studies to substantiate the project of inter-basin redistribution of the flow of Siberian rivers, *Tr. Leningr. Politekh. Inst.,* No. **346,** (1976).

Калачев, Н. С., Лаврентьева, Л. Д. и Казыбаев, А. К. Некоторые аспекты обводнения Казахстана и сопредельных с ним территорий водами сибирских рек, Сб. *Проблемы гидро энергетики и водного хозяйства,* вып. 8. (1971).

Kalachev, N. S., Lavrenteva, L. D. and Kazybaev, A. K. Some aspects of irrigation of Kazakhstan and adjacent areas with the waters of the Siberian Rivers, *Probl. Gidroenerg. Vodn. Khoz.,* issue 8, (1971).

Калачев, Н. С., Лаврентьева, Л. Д. и Павленко, В. П. Аргут–Катунь–Бухтарминский энерговодохозяйственный комплекс, Сб. *Проблемы гидро энергетики и водного хозяйства,* вып. 9. (1972).

Kalachev, N. S., Lavrenteva, L. D. and Pavlenko, V. P. The Argut–Katun–Bukhtarma irrigation and hydraulic power complex, *Probl. Gidroenerg. Vodn. Khoz.,* issue 9, (1972).

Кореньков, О. Б. Об использовании стока реки Волги для широкого развития орошения Нижнего Дона и Северного Кавказа, Сб. *Географические проблемы изучения и освоения природных ресурсов Нижнего Дона и Северного Кавказа,* Ростов-на-Дону, (1971).

Korenkov, O. B. The utilization of the Volga runoff for extensive development of irrigation in the Lower Don Area and in Northern Caucasus, *Geograficheskiye Problemy Izucheniya i Osvoeniya Prirodnykh Resursov Nizhnego Dona i Severnogo Kavkaza,* Rostov-na-Donu, (1971).

Кореньков, О. Б. О переброске в нижнее течение р. Дон части стока р. Волги для нужд народного хозяйства Ростовской области, Книга *"Совершенствование проектирования мелиорации и водохозяйственных объектов на Северном Кавказе",* Ростов-на-Дону, (1973).

Korenkov, O. B. On partial diversion of the Volga runoff to the Lower Don River to meet the economic requirements of the Rostov Region, *Sovershenstvovaniye Proektirovaniya Melioratsii i Vodokhozyaystvennykh Obektov na Severnom Kavkaze,* Rostov-na-Donu, (1973).

Кошелев, А. П. К истории вопроса переброски стока рек в России, Сб. *Природные условия Западной Сибири и переброска стока рек в Среднюю Азию,* Новосибирск, (1975).

Koshelev, A. P. The history of streamflow rechannelling in Russia, *Prirodnye Usloviya Zapadnoy Sibiri i Perebroska Stoka rek v Srednyuyu Aziyu, Novosibirsk,* (1975).

Лисицина, И. Н. и Ронин, В. Л. Комплексное использование водных ресурсов Волги, *Гидротехника и мелиорация,* № 11, (1973).

Lisitsina, I. N. and Ronin, V. L. Integrated utilization of the Volga water resources, *Gidrotekh. Melior.,* No. 11, (1973).

Малик, Л. К. Нужно ли перебрасывать воды западносиоирских рек на юг, *Природа,* № 6, (1973).

Malik, L. K. Is there any need for the water of the west Siberian Rivers to be diverted southward? *Priroda,* No. 6, (1973).

Малинина, Т. И. и Распопов, И. М. *Озёра Лача, Воже, Кубенское как потенциальные ресурсы пополнения вод реки Волги,* Материалы Всесоюзной научной конференции по проблемам комплексного использования и охраны водных ресурсов бассейна Волги, вып. 1, Пермь, (1975).

Malinina, T. I. and Raspopov, I. M. *Lakes Lacha, Vozhe and Kubenskoe as potential resources to replenish the Volga Water,* Materialy Vsesoyuznoy Nauchnoy Konferentsii po Probleman Kompleksnogo Ispolzovaniya i Okhrany Vodnykh Resursov Basseyna Volgi, Vol. 1, Perm, (1975).

Мошков, А. Б. и Симаков, Г. В. Перспективы новых воднотранспортных соединений в связи с региональной переброской стока рек, *Труды Ленинградского Политехнического института,* № **351,** (1976).

Moshkov, A. B. and Simakov, G. V. The prospects of new water transport communications in connection with a streamflow diversion, *Tr. Lening. Politekh. Inst.,* No. **351,** (1976).

Перераспределение речных ресурсов, *Народное хозяйство Казахстана,* № 1, (1977).

The redistribution of river resources, *Nar. Khoz. Kazakhstana,* No. 1, (1977).

Природные условия Западной Сибири и переброска стока рек в Среднюю Азию, Новосибирск, (1975).
The Environment of West Siberia and the Diversion of the River Flow to Central Asia, Novosibirsk, (1975).

Сакс, В. Н., Широков, В. М. и Яншин, А. Л. Задачи научных исследований в Сибири при разработке вопросов переброски части стока рек в Среднюю Азию, Сб. *Природные условия Западной Сибири и переброска стока рек в Среднюю Азию,* Новосибирск, (1975).

Saks, V. N., Shirokov, V. M. and Yanshin, A. L. The objectives of research in Siberia to develop the problems of the partial diversion of the river flow to Central Asia, *Prirodnye Usloviya Zapadnoy Sibiri i Perebroska Stoka rek v Srednyuyu Aziyu,* Novosibirsk, (1975).

Саруханов, Г. Л. Переброска стока северных рек, *Труды Гидропроекта,* сб. 16, (1969).

Sarukhanov, G. L. Diversion of run-off of the northern rivers, *Tr. Gidroproekta,* issue 16, (1969).

Саруханов, Г. Л. *К переброске стока северных рек по Волжскому каскаду,* Вторая конференция по изучению водоёмов бассейна Волги, "Волга-2"; Борок, (1974).

Sarukhanov, G. L. *The diversion of the flow of the Northern Rivers through the Volga Cascade,* Vtoraya Konferentsiya po Izucheniyu Vodoemov Basseyna Volgi, Volga-2; Borok, (1974).

Саруханов, Г. Л. Переброска стока северных рек в Волгу, *Речной транспорт,* № 7, (1975).

Sarukhanov, G. L. The Diversion of the Northern Rivers Flow to the Volga, *Rechn. Transp.* No. 7, (1975).

Чокин, Ш. Ч. Проблемы водного благоустройства Срединного региона СССР. *Водные ресурсы,* № 6, (1974).

Chokin, Sh. Ch. The problems of water supply improvement in the central region of the USSR, *Vodn. Resur.* No. 6, (1974).

Чокин, Ш. Ч. и Калачев, Н. С. *Реки Сибири потекут в Казахстан и Среднюю Азию,* Алма-Ата, (1974).

Chokin, Sh. Ch. and Kalachev, N. S. *The Rivers of Siberia will Flow to Kazakhstan and Central Asia,* Alma-Ata, (1974).

Чокин, Ш. Ч., Захаров, В. П., Калачев, Н. С. и Киктенко, В. А. Территориальное перераспределение водных ресурсов и задачи гидрологических исследований в Срединном регионе СССР. *Труды IУ Весоюзного Гидрологического Съезда,* том **4**, Ленинград (1976).

Chokin, Sh. Ch., Zakharov, V. P., Kalachev, N. S. and Kiktenko, V. A. The large-scale areal redistribution of water resources and the objectives of hydrological research in the USSR Central Region *Tr. IV Vses. Gidrol St"êzda,* **4,** Leningrad (1976).

Шимельмиц, И. Я., Бостанджогло, А. А., Грин, Г. Б. и Ауце, Д. А. Некоторые аспекты проблемы изъятия стока Сибирских рек для водообеспечения Средней Азии и Казахстана, Сб. *Влияние перераспределения стока рек бассейна Оби на природу тайги Срединного регноиа,* Иркутск, (1975).

Shimelmits, I. Ya., Bostandzhoglo, A. A., Grin, G. B. and Autse, D. A. Some aspects of the problem of the partial withdrawal of the flow of the Siberian rivers for the water supply of Central Asia and Kazakhstan, *Vliyaniye Pereraspredeleniya Stoka rek Basseyna Obi na Prirodu Taygi Sredinnogo Regiona,* Irkutsk, (1975).

ECONOMIC ASPECTS

Castle, Emery, N. *The economics of massive inter-basin water transfers,* Seminar on the Continental Use of Arctic-Flowing Rivers, Washington State Univ., pp. 153–163 (1968).

Crutchfield, James A. NAWAPA: A continental water scheme (symposium): Economic considerations, *Bull. atom. Scient.,* **23,** 17–21 (September 1967).

Cummings, Ronald G. *Inter-basin Water Transfers: A Case Study in Mexico,* Published by Resources for the Future Inc., Distributed by Johns Hopkins Press, Baltimore and London, (1974).

Easter, K. William, *Inter-basin water transfers—economic issues and impacts,* Proc. 4th Am. Water Resources Conf., pp. 191–200 (1968).

Eckart, N. A. Benefits accruing from the Hetch Hetchy Project, San Francisco water supply, *J. Am. Wat. Wks Assn.,* **28,** 1211–1231 (1936).

Hartman, L. M. *Water resources and economic development of the west,* Report No. 14, Conf. Proceedings of the Committee on the Economics of Water Resources Development of the Western Agric. Economics Research Council, Berkeley, CA, (1965).

Hartman, L. M. and Seastone, D. A. Efficiency criteria for market transfers of water, *Water Resour. Res.,* **1** (2), 165–71 (1965).

Hartman, L. M. and Seastone, D. A. Regional economic interdependencies and water use, in Kneese and Smith, *Water Research,* Johns Hopkins Press, Baltimore, (1966) pp. 215–231.

Hartman, L. M. and Seastone, Don, *Water Transfers: Economic Efficiency and Alternative Institutions,* Published for Resources for the Future Inc. by Johns Hopkins Press, Baltimore and London, (1970).

Heinly, Burt A. Surplus waters of the Los Angeles Aqueduct, *Engng News, N.Y.,* **69,** 506–510, Edit: **70,** 32 (1913).

Howe, Charles W. *Economics of large-scale transfers,* AAAS Symposium on Arid Lands, Dallas, Texas, pp. 371–382, processed, (30 December 1968).

Howe, Charles W. *Economic issues related to large-scale water transfers in the U.S.A.,* IIASA Task Force Meeting on Interregional Water Transfers, Laxenburg, (11–14 October, 1977).

Howe, Charles W. and Easter, K. William, *Inter-basin Transfers of Water: Economic Issues and Impacts,* Johns Hopkins Press, Baltimore, (1971).

Kuiper, E. and Murray, J., *Water supplies for North American food production,* 29 pp. Proc., (1967).

Linaweaver, F. P. Jr and Clark, C. Scott, Costs of water transmission, *J. Am. Wat. Wks Ass.,* **56,** 1549–1560 (1964).

Millham, C. B. and Russell, R. A. On the economic impact of large diversions of Snake River Waters, *Water Resour. Bull.,* **7,** 925–934 (1971).

Schramm, Gunter, Regional versus inter-regional efficiency in resource allocations, *Ann. Reg. Sci.* **4,** 1–14 (December 1970).

Sorensen, J. N. and Clark, R. T. Marginal value of irrigation water and a case study of transfer in South-eastern Wyoming, *Wyo. Agric. Exp. Stn. Bull.,* No. 511, 42 pp. (1970).

Stoevener, Herbert H. *The economics of water diversion,* Symposium on Northwest–Southwest Water Diversion Issues, Oregon State Univ. pp. 47–57 (1967).

Young, Robert A. and Martin, William E. The economics of Arizona's water problem, *Ariz. Rev.,* pp. 9–18 (16 March 1967).

Воробьев, Б. В. Влияние создания водохранилищ при переброске стока северных рек и озер в Волгу на сельское хозяйство, *Труды координационных совещаний по гидротехнике,* вып. 83, (1973).

Vorobyev, B. V. The effect of newly constructed water reservoirs on agriculture when diverting the water of the northern rivers and lakes to the Volga, *Tr. Koord. Sovesch. Gidrotekh.,* issue 83, (1973).

Пучкова, Н. В. Народохозяйственные предпосылки территориального перераспределения стока, *Труды Гидропроекта,* сб. **29,** (1973).

Puchkova, N. V. Economic prerequisites for large-scale streamflow redistribution, *Tr. Gidroproekta,* issue 29, (1973).

Сергин, С. Я. К целесообразности переброски Сибирских вод в Среднюю Азию, Сб. *Природные условия Западной Сибири и переброска стока рек в Среднюю Азию,* Новосибирск, (1975).

Sergin, S. Ya. Economic soundness of the diversion of Siberia's river waters to Central Asia, *Prirodnye Usloviya Zapadnoy Sibiri i Perebroska Stoka rek v Srednyuyu Aziyu,* Novosibirsk, (1975).

Хахин, Н. А. и Шатаева, С. Г. Оценка влияния на водный транспорт изъятия части стока Оби и Иртыша при переброске его в бассейн Аральского моря, Сб. *Оценка возможных изменений режима низовьев и устьев рек Арктической зоны Западной Сибири под влиянием водохозяйственных мероприятий,* Ленинград, (1976).

Khakhin, N. A. and Shataeva, S. G. Evaluating the effect of the partial withdrawal of the flow of the Ob and the Irtysh Rivers on water transport when conveying this part of the flow to the Aral Sea Basin, *Otsenka Vozmozhnykh Izmeneniy Rezhima Nizovyev i Ustiev rek Arkticheskoy Zony Zapadnoy Sibiri pod Vliyaniem Vodokhozyaystvennykh Meropriyatiy,* Leningrad, (1976).

ENVIRONMENTAL ISSUES

Berger, Bernard B. Social and ecological implications of proposals for out-of-basin river diversions, *Proc. Am. Soc. civ. Engrs,* **97,** 95–110 (1971).

Hagan, R. M. and Roberts, E. B. Ecological impacts of water projects in California, *Proc. Am. Soc. civ. Engrs, J. Irrig. Drain. Div.,* **98,** (IRI), 25–48 (1972).

McGauhey, Percy H. Physical implications of large-scale water transfers, in McGinnies and Goldman, *Arid Lands in Perspective,* U. of Arizona Press, (1968) pp. 358–363.

Nace, Raymond, Water in flatland, *Geo. Sci. News,* **1** (1), 2–5, 18–22 (1967).

Ortolano, Leonard, *Environmental assessments in water resources planning,* IIASA Task Force Meeting on Interregional Water Transfers, Laxenburg, (11–14 October 1977).

Radforth, Norman V. NAWAPA and Muskeg, in Royal Soc. of Canada, *Water Resources of Canada,* Univ. of Toronto Press, pp. 27–28, (1967).

Thomas, Gerald W. and Box, Thadis W. *Social and ecological implications of water importation into arid lands,* AAAS Symposium on Arid Lands, Dallas, Texas, pp. 363–374, processed, (1968).

Tinney, E. Roy, NAWAPA: A continental water scheme, Symposium on Engineering Aspects, *Bull. atom. Scient.,* **23,** 21–25 (September 1967).

Антонов, В. С. Многолетняя изменчвость стока—экологический критерий оценки допустимого изъятия вод из Северных и Сибриских рек, Сб. *Оценка возможных изменений режима низовьев и устьеы рек Арктической зоны Западной Сибири под влиянием водохозяйственных мероприятий,* Ленинград, (1976).

Antonov, V. S. Normal streamflow variation as an ecological criterion for assessing the permissible level of water withdrawal from Northern and Siberian rivers, *Otsenka Vozmozhnykh Izmeneniy Nizovyev i Ustyev rek Arkticheskoy Zony Zapadnoy Sibiri pod Vliyaniem Vodokhozyaystvennykh Meropriyatiy,* Leningrad, (1976).

Батраков, Ю. В. Изменения направления стока Сибирских рек и вращение Земли, *Астрономический журнал,* Т. **48,** № 5, (1971).

Batrakov, Yu. V. Variation in the direction of the flow of the Siberian rivers and the rotation of the earth, *Astr. Zh.* **48,** No. 5, (1971).

Бачурин, Г.В., Ильина, И.С., Михайлов, Ю.П., Петров, И.Б. и Широков, В.М. О влиянии переброски стока Сибирских рек на природу таежных геосистем, Сб. *Влияние перераспределения стока рек бассейна Оби на природу тайги Срединного региона,* Иркутск, (1975).

Bachurin, G. V., Ilina, I. S., Mikhaylov, Yu. P., Petrov, I. B. and Shirokov, V. M. The influence of the diversion of Siberian river flow on the environmental conditions of the taiga geosystems, *Vliyaniye Pereraspredeleniya Stoka rek Basseyna Obi na Prirodu Taygi Sredinnogo Regiona,* Irkutsk, (1975).

Братцев, А. П. Возможные изменения природных условий в нижнем течении Оби при переброске части ее стока в бассейн Волги, Сб. *Оценка возможных изменений режима низовьев и устьев рек Арктической зоны Западной Сибири под влиянием водохозяйственных мероприятий,* Ленинград, (1976).

Bratsev, A. P. Possible environmental changes in the lower reaches of the Ob River when part of its flow is diverted to the Volga Basin, *Otsenka Vozmozhnykh Izmeneniy Rezhima Nizovyev i Ustyev rek Arkticheskoy Zony Zapadnoy Sibiri pod Vliyaniem Vodokhozyaystvennykh Meropriyatiy,* Leningrad, (1976).

Братцев, А. П. Влияние переброски стока р. Печоры в бассейн р. Волги на природные условия и народное хозяйтсво Коми АССР, *Серия научных докладов, АН СССР Коми Филиал,* вып. 26, (1976).

Bratsev, A. P. The influence of the diversion of the flow of the Pechora River to the Volga Basin on the environment and economy of the Komi ASSR, *Ser. Neuchnykh Dokl., AN SSSR Komi Fil.* issue 26, (1976).

Вендров, С. Л. Географические аспекты проблемы территориального перераспределения речного стока на Русской равнине, Сб. *Природные ресурсы Русской равнины и перспектива их использования,* Москва, (1974).

Vendrov, S. L. Geographical aspects of the problem of large-scale areal redistribution of streamflow on the Russian Plain, *Prirodnye Resursy Russkoy Ravniny i Perspektiva ikh ispolzovaniya,* Moskva, (1974).

Вендров, С. Л. О некоторых задачах географических наук в связи с проблемами водохозяйственного благоустройства, *Вопросы географии,* сб. **100,** (1976).

Vendrov, S. L. Some of the tasks facing geographical sciences in connection with the problems of water-resources development *Vop. Geogr.* **100,** (1976).

Вильнер, М. Я. Градостроительный аспект переброски части стока Сибирских рек в бассейн Аральского моря, Сб. *Опыт разработки и реализации схем и проектов районной планировки,* Киев, (1976).

Vilner, M. Ya. The urban aspect of the partial transfer of the flow of the Siberian rivers to the Aral Sea Basin, *Opyt Razrabotki i Realizatsii Skhem i Proektov Rayonnoy Planirovki,* Kiev, (1976).

Влияние межбассейнового перераспределения речного стока на природные условия Европейской территории и Срединного региона СССР, *Тезисы докладов Совещания,* Москва (1975).

The effects of inter-basin river flow redistribution on the environmental conditions of the european area and the centıal region of the USSR, *Tezisy Dokl. Soveshchaniya,* Moskva (1975).

Влияние перераспределения стока рек бассейна Оби на природу тайги Срединного региона, *Материалы совещания,* Иркутск (1975).

The effect of the streamflow redistribution of the Ob River on the tayga environment of the central region, *Mater. Soveshchaniya,* Irkutsk (1975).

Волков, И. А. Ландшафтные и геологогеоморфологические предпосылки решения проблемы переброски части стока Сибирских рек на юг, Сб. *Природно-мелиоративная характеристика Средней Азии и Казахстана,* Пущино, (1976).

Volkov, I. A. Environmental geological and geomorphological factors of the partial diversion of the Siberian rivers run-off to the south, *Prirodnomeliorativnaya Kharakteristika Sredney Azii i Kazakhstana,* Puschchino, (1976).

Воробьев, Б. В. Взаимосвязь экологических и экономических процессов при освоении водных ресурсов рек и озер, *Труды координационного совещания по гидротехнике,* вып. 102, (1975).

Vorobyev, B. V. The interaction of ecological and economic processes in developing the water resources of rivers and lakes, *Tr. Koord. Soveshch. Gidrotekh.,* issue 102, (1975).

Гаврилова, Л. А. Особенности метеорологического режима в низовьях и устьевой области реки Оби и пути оценки его возможных изменений под влиянием водохозяйственных мероприятий, Сб. *Оценка возможных изменений режима низовьев и устьев рек Арктической зоны Западной Сибири под влиянием водохозяйственных мероприятий,* Ланинград, (1976).

Gavrilova, L. A. Meteorological conditions in the lower and mouth areas of the Ob River and the methods of evaluating possible variations of these conditions due to water management activities, *Otsenka Vozmozhnykh Izmeneniy Rezhima Nizovyev i Ustyev rek Arkticheskoy Zony Zapadnoy Sibiri pod Vliyaniyem Vodokhozyaystvennykh Meropriyatiy,* Leningrad, (1976).

Герасимов, И. П. и Гиндин, А. М. *Проблемы переброски сибирских речных вод на пустынные равнины Средней Азии и Казахстана,* XXIII Международный Географический конгресс, Симпозиум "Человек и среда", Москва, (1976).

Gerasimov, I. P. and Gindin, A. M. *The problems of diversion of the Siberian river waters to the desert plains of Central Asia and Kazakhstan,* XXIII Mezhdunarodnyy Geograficheskii Kongress, Simpozium: Chelovek i Sreda, Moskva, (1976).

Городецкая, М. Е. Геоморфологические условия переброски части стока западносибирских рек в бассейн Сырдарьи, Сб. *Природно-мелиоративная характеристика Средней Азии и Казахстана,* Пущино, (1976).

Gorodetskaya, M. E. The geomorphological conditions of the partial diversion of the flow of the West Siberian rivers to the Syr-Darya River Basin, *Prirodno-meliorativnaya Kharakteristika Sredney Azii i Kazakhstana,* Pushchino, (1976).

Дмитриев, В. С., Пославский, В. В. и Решёткина, Н. М. Мелиоративные прогнозы – основа выбора технических решений при переброске части стока северных рек на юг, *Водные ресурсы,* № 5, (1977).

Dmitriev, V. S., Poslavskii, V. V. and Reshetkina, N. M. Reclamation forecasts as the basis to selecting technological schemes to transfer some part of runoff of northern rivers to the south, *Vodn. Resur.,* No. 5, (1977).

Дроздов, О. А., Сорочан, О. Г. и Шикломанов, И. А. Предварительная оценка возможных изменений глобального влагооборота под влиянием хозяйственной деятельности, *Водные ресурсы,* № 6, (1976).

Drozdov, O. A., Sorochan, O. G. and Shiklomanov, I. A. Preliminary estimation of probable changes in global water circulation under the effect of economic activities, *Vodn. Resur.* No. 6, (1976).

Зарубаев, Н. В. Вопросы охраны окружающей среды в зонах межбассейнового перераспределения речного стока, *Труды Ленинградского Политехничесного института,* № **351,** (1976).

Zarubaev, N. V. Environmental protection problems in inter-basin river flow redistribution zones, *Tr. Leningr. Politekh. Inst.,* No. **351,** (1976).

Катков, Н. С. и Новицкий, О. П. Рыбное хозяйство Обь–Иртышского бассейна в условиях отбора части стока для переброски в бассейн Аральского моря. Сб. *Оценка возможных изменений режима низовьев и устьев рек Арктической зоны Западной Сибири под вкиянием водохозяйственных мероприятий,* Ленинград, (1976).

Katkov, N. S. and Novitskiy, O. P. Fishing practices in the Ob–Irtysh Basin with part of the flow removed for conveyance to the Aral Sea Basin, *Otsenka Vozmozhnykh Izmeneniy Rezhima Nizobyev i Ustiev rek Arkticheskoy Zony Zapadnoy Sibiri pod Vliyaniem Vodokhozyaystvennykh Meropriyatiy,* Leningrad, (1976).

Кесь, А. С. и Грязнова, Т. П. Геоморфологические условия зоны переброски части стока Сибирских рек на междуречье низовий Сырдарьи – Амударьи, Сб. *Природно-мелиоративная характеристика Средней Азии и Казахстана,* Пущино, (1976).

Kes, A. S. and Gryaznova, T. P. Regional geomorphological conditions divert part of the flow of the Siberian Rivers to the Syrdarya–Amudarya lower reaches interfluve area, *Prirodnomeliorativnaya Kharakteristika Sredney Azii i Kazakhstana,* Pushchino, (1976).

Макеев, О. В. и Чигир, В. Г. Переброска части речного стока северных рек на юг и её возможное влияние на мерзлотные условия в лесной зоне Западной Сибири, Сб. *Влияние перераспределения стока рек бассейна Оби на природу тайги Срединного региона,* Иркутск (1975).

Makeev, O. V. and Chigir, V. G. The partial diversion of the flow of the northern rivers southward and its possible effect on the permafrost in the forest zone of West Siberia, *Vliyaniye Pereraspredeleniya Stoka rek Basseyna Obi na Prirodu Taygi Sredinnogo Regiona,* Irkutsk, (1975).

Мартынов, А. В. К оценке изменения природных условий в результате переброски части стока сибирских рек, Сб. *Изучение природы, хозяйства и населения Сибири,* Иркутск, (1975).

Martynov, A. V. Evaluating the environmental changes due to the partial diversion of the flow of the Siberian rivers, *Izucheniye Prirody, Khozyaystva i Naseleniya Sibiri,* Irkutsk, (1975).

Мелиорация почв Средней Азии, Казахстана и Западной Сибири в связи с переброской части стока Сибирских рек в южные районы страны, *Тезисы докладов совещания,* Пущино (1973).

Soil reclamation in Central Asia, Kazakhstan and West Siberia in connection with the partial diversion of the flow of the siberian rivers to the country's southern Areas, *Tezisy Dokl. Soveshchaniya,* Pushchino, (1973).

Михайлов, Н. И. Проблемы перераспределения стока рек Срединного района СССР, *Вестник МГУ, серия У – География,* № 6, (1974).

Mikhaylov, N. I. The problems of river flow redistribution in the USSR central region, *Vest. mosk. gos. Univ., ser. V-Geogr.,* No. 6, (1974).

Михайлов, Н. И. Естественные тенденции развития ландшафтов Западной Сибири и возможные изменения их в связи с проблемами перераспределелия стока, Сб. *География и хозяйство Красноярского края,* Красноярск, (1975).

Mikhaylov, N. I. Natural landscape development trends in West Siberia and their possible changes in connection with the problems of streamflow reallocation, *Geografiya i Khozyaystvo Krasnoyarskogo Kraya,* Krasnoyarsk, (1975).

Михайлов, Н. И., Криволуцкий, А. Е., Николаев, В. А., Скорняков, В. А. и Тимашев, И. Е. *Прогноз физико-географических следствий переброски части стока Сибирских рек на юг,* XXIII Международный Географический конгресс, Секция "Общая физическая география", Москва, (1976).

Mikhaylov, N. I., Krivolutskiy, A. E., Nikolaev, V. A., Skornyakov, V. A. and Timashev, I. E. *Forecasting the environmental consequences of the partial diversion of the flow of the Siberian rivers to the south,* XXIII Mezhdunarodnyi Geograficheskiy Kongress, Sektsiya, Obshchaya Fizicheskaya Geografiya, Moskva, (1976).

Михайлов, Ю. П. Влияние переброски речного стока северных рек на природу таёжных геосистем, *Известия Всесоюзного Географического общества,* т. **107**, вып. 1, (1975).

Mikhaylov, Yu. P. The influence of the diversion of the northern rivers flow on the taiga geosystems, *Izv. Vses. Geogr. Obshchestva,* **107,** (1), (1975).

Михайлов, Ю. П. Географические аспекты охраны природы Обского Севера в связи с перераспределением стока Оби и Иртыша, Сб. *Влияние перераспределения стока рек бассейна Оби на природу тайги Срединного региона,* Иркутск, (1975).

Mikhaylov, Yu. P. Geographical aspects of the environmental protection of the Ob northern area in connection with the redistribution of the flow of the Ob and the Irtysh, *Vliyaniye Pereaspredeleniya Stoka rek Basseyna Obi na Prirodu Taygi Sredinnogo Regiona,* Irkutsk, (1975).

Никонов, С. П. Перспективы развития рыбного хозяйства Обь–Иртышского бассейна при условии отбора части его стока, Сб. *Влияние перераспределения стока рек бассейна Оби на природу тайги Срединного региона,* Иркутск, (1975).

Nikonov, S. P. The prospects of development of the fishing in the Ob–Irtysh Basin with part of its flow withdrawn, *Vliyaniye Pererapredeleniya Stoka rek Basseina Obi na Prirodu Taygi Sredinnogo Regiona,* Irkutsk, (1975).

Оценка возможных изменений режима низовьев и устьев рек Арктической зоны Западной Сибири под влиянием водохозяйственных мероприятий, Тезисы Докладов, Ленинград (1976).

Evaluating possible variations of the regime in the lower and mouth reaches of the arctic rivers of west Siberia under the influence of water management activities, *Tezisy Dokl.* Leningrad (1976).

Ретеюм, А. Ю. *Географические вопросы проектирования и прогнозирования в разработке планов переброски стока рек Европейского Севера,* Материалы III Всесоюзного совещания по прикладной географии, Иркутск, (1975).

Reteyum, A. Yu. *Geographical questions of planning and forecasting in working out schemes of streamflow diversion of the rivers of the northern European part of the USSR,* Materialy III Vsesoyuznogo Soveshchaniya po Prikladnoy Geografii, Irkytsk, (1975).

Степанов, И. Н. и Озерова, Е. Н. Возможные изменения природно-мелиоративных условий Срединного региона СССР в связи с переброской стока Сибирских рек на юг страны, Сб. *Природно-мелиоративная характеристика Средней Азии и Казахстана,* Пущино, (1976).

Stepanov, I. N. and Ozerova, E. N. Possible changes in the environmental and meliorative conditions of the USSR Central Region in connection with the diversion of the flow of the Siberian rivers to the south, *Prirodnomeliorativnaya Kharakteristika Sredney Azii i Kazakhstana,* Pushchino, (1976).

Трешников, А. Ф. Задачи изучения природы Арктики в связи с частичныи изъятием стока северных и сибирских рек, Сб. *Оценка возможных изменений режима низовьев и устьев рек Арктической зоны Западной Сибири под влиянием водохозяйственных мероприятий,* Ленинград, (1976).

Treshnikov, A. F. Objectives for the study of the arctic environment in connection with partical withdrawal of runoff from northern and Siberian rivers, *Otsenka Vozmozhnykh Izmeneniy Rezhima Nizovyev i Ustiev rek Arkticheskoy Zony Zapadnoy Sibiri pod Vliyaniem Vodokhozyaystvennykh Meropriyatiy,* Leningrad, (1976).

Ходжибаев, Н. Н. *Прогноз изменения гидрогеолого–мелиоративных условий территории Средней Азии, затрагиваемых межбассейновыми перебросками рек,* Международный Геологический конгресс, ХХУ сессия, Москва, (1976).

Khodzhibaev, N. N. *Forecasting the changes of hydrogeological and reclamative conditions in the area of Central Asia involved in inter-basin streamflow transfer,* Mezhdunarodnyi Geologicheskiy Kongress, XXV Sessiya, Moskva, (1976).

Ходжибаев, Н. Н., Арипов, К. М., Чеботарёва, Г. Б., Магднев, В. У. и Рахимов, М. Р. К вопросу изменения гидрогеолого–мелиоративных условий территории Средней Азии в связи с межбассейновыми перебросками стока, Сб. *Вопросы изучения и прогноза режима и баланса грунтовых вод орошаемых земель аридной зоны СССР,* Ташкент, (1975).

Khodzibaev, N. N., Aripov, K. M., Chebotareva, G. B., Magdnev, V. U. and Rakhimov, M. R. Changes in the hydrogeological and reclamative conditions of the area of Central Asia in connection with inter-basin streamflow transfer, *Voprosy Izucheniya i Prognoza Rezhima i Balansa Gruntovykh vod Orashaemykh Zemel Aridnoy Zony SSSR,* Tashkent, (1975).

HYDROLOGICAL ASPECTS

Ackerman, William C. *Water transfers—possible de-eutrophication of the Great Lakes,* Natural Research Council, 25 pp. Proc., (11 March 1968).

Dutton, R. D. and Millham, C. B. On large diversions from the Northwest–normal and high flow years, *Water Resour. Bull.,* **9,** 231–242 (1973).

Kirshner, Louis D. *Effects of diversions on the Great Lakes,* Proc. Great Lakes Water Resources Conf., Paper R-4, pp. 277–310 (1968).

Laing, Arthur, *Great Lakes Water Levels,* Brochure: Dept of Northern Affairs and National Resources, Water Resources Branch, 14 pp. (1964).

Meserow, Albert J. *Great Lakes water—Is there enough?* Proc. 1st Am. Water Resources Ass. Conf., pp. 365–371 (1965).

Odom, Leo M. Atchafalaya diversion and its effect on the Mississippi River, *Trans. Am. Soc. civ. Engrs,* **116,** 503–547 (1951).

Salisbury, E. F. Influence of diversion on the Mississippi and Archafalaya Rivers, *Trans. Am. Soc. civ. Engrs,* **102,** 75–103, Disc: Odom, Leo M., Lane, E. W. and Salisbury, E. F. **102,** 103–122 (1937).

White, W. A., Tischler, L. F. and Austin, T. A. Water quality prediction within an inter-basin transfer system, *Water Resour. Bull.,* **8** (3), (June 1972).

Yantis, Hugh C. *Water quality considerations,* Symposium on Aspects of the Diversion of Mississippi River Water to Texas and New Mexico, pp. 30–37 (1969).

Антонов, В. С. К проблеме безвозвратного изъятия части стока северных рек, *Труды Арктического и Антарктического научно-исследовательского института,* т. **323,** Ленинград (1976).

Antonov, V. S. The problem of irretrievable removal of part of the flow of the northern rivers, *Tr. arkt. antark. nauchno-Issled. Inst.,* **323,** Leningrad (1976).

Архипова, Е. Г. Современный и перспективный водный и солевой баланс и возможные изменения гидрологического режима Каспийского моря, *Труды ВНИИ морского рыбного хозяйства и океанографии,* вып. **108,** (1975).

Arhipova, E. G. Present and expected water and salt balance and possible changes of hydrologic regimen of the Caspian Sea, *Tr. vses. nauchno-issled. Inst. morsk. ryb. Khoz. Okaenogr.,* **108,** (1975).

Баранов, В. А., Бочков, Н. М. и Мясникова, Е. В. Основные черты прогноза изменения качества воды в бассейнах р. Волги и Аральского моря при переброске в них части стока северных и сибирских рек, *Труды IУ Всесоюзного Гидрологического съезда,* т. **9,** Ленинград (1976).

Baranov, V. A., Bochkov, N. M. and Myasnikova, E. V. The basic characteristics of prediction of water quality variations in the basins of the Volga and the Aral Sea on diverting part of the flow of the northern and Siberian rivers, *Tr. IV Vses. Gidrol. S"ëzda,* **9,** Leningrad (1976).

Бейром, С. Г. и Широков, В. М. Предполагаемые изменения гидрологических и гидрогеологических условий в земельном фонде Сибири при переброске стока рек в Среднюю Азию, *Земельные ресурсы Сибири,* Новосибирск (1975).

Beyrom, S. G. and Shirokov, V. M. Likely changes in the hydrological and hydrogeological conditions of lands of Siberia on diversion of river flow to Central Asia, *Zemelnye Resur. Sibiri,* Novosibirsk (1975).

Бронфман, А. М. Некоторые океанологические и водохозяйственные аспекты компенсирующей переброски стока в бассейн Азовского моря, *Географические исследования на Северном Кавказе и Нижнем Дону,* Ростов-на Дону, (1973).

Bronfman, A. M. Some oceanological and water management aspects of the compensating streamflow diversion to the Basin of the Azov Sea, *Geograficheskiye Issledovaniya na Severnom Kavkaze i Nizhnem Donu,* Rostov-na-Donu, (1973).

Вельнер, Х. А. и Каск, А. Г. Об определении величины остаточного расхода воды при переброске стока, *Труды Всесеюзного научно-технического совещания "Разработка и организация водоохранных мероприятий",* Харьков, (1973).

Velner, Kh. A. and Kask, A. G. Determination of the residual water discharge rate in case of streamflow transfer, *Trudy Vsesoyuznogo Nauchno-Tekhnicheskogo Soveshchaniya "Razrabotka i Organizatsiya Vodookhrannykh Meropriyatiy",* Kharkov, (1973).

Вендров, С. Л. Проблемы территориального перераспределения речного стока, *Изв. АН СССР, серия Географ,* № 1, (1975).

Vendrov, S. L. The problems of large-scale areal redistribution of river flow, *Izv. AN SSSR, Ser. Geogr.* No. 1, (1975).

Вендров, С. Л. Использование и преобразование рек, Сб. *Природные ресурсы Русской равнины в прошлом, настоящем и будущем,* Москва, (1976).

Vendrov, S. L. The utilization and remaking of rivers, *Prirodnye Resursy Russkoy Ravniny v Proshlom, Nastoyashchem i Budushchem,* Moskva, (1976).

Вендров, С. Л. и Комлев, А. М. Состояние и перспективы иучения водных ресурсов Западной Сибири в связи с их народохозяйственным использованием, *Труды Западно-Сибирского регионального научно-исследовательского гидрометеорологического института,* вып. 26, (1976).

Vendrov, S. L. and Komlev, A. M. The state and prospects of water resources research for West Siberia in connection with water resources economic utilization, *Tr. Zapadno-Sib. Reg. Nauchno-Issled. Gidrometeorol. Inst.,* issue 26, (1976).

Герарди, И. А., Харченко, С. И. и Роо, С. С. Гидрологические проблемы переброски части стока северных и сибирских рек на южный склон СССР, *Труды IУ Всесоюзного Гидрологического съезда,* т. **4,** Ленинград (1976).

Gerardi, I. A., Kharchenko, S. I. and Roo, S. S. Hydrological problems of the partial diversion of the flow of the Northern and Siberian Rivers to the Southern Slope of the USSR, *Tr. IV Vses. Gidrol. S"êzda,* **4,** Leningrad (1976).

Гришанин, К. В. Возможные изменения режима низовьев Оби при частичном отъёме стока, Сб. *Оценка возможных изменений режима низовьев и устьев рек Арктической зоны Западной Сибири под влием водохозяйственных мероприятий,* Ленинград, (1976).

Grishanin, K. V. Possible changes in the regimen of the lower reaches of the Ob River in case of partial flow withdrawal, *Otsenka Vozmozhnykh Izmeneniy Rezhima Nizovyov i Ustyev rek Arkticheskoy Zony Zapadnoy Sibiri pod Vliyaniem Vodokhozyaystvennoykh Meropyatiy,* Leningrad, (1976).

Донченко, Р. В. Оценка изменеий характеристик ледового режима рек Оби и Иртыша в связи с переброской их стока, Сб. *Оценка возможных изменений режима низовьев и устьев рек Арктической зоны Западной Сибири под влиянием водохозяйственных мероприятий,* Ленинград, (1976).

Donchenko, R. V. Evaluating the changes in the characteristics of the ice conditions of the Ob and the Irtysh Rivers in connection with the transfer of their flow, *Otsenka Vozmozhnykh Izmeneniy Rezhima Nizovyev i Ustyev rek Arkticheskoy Zony Zapadnoy Sibiri pod Vliyaniem Vodokhozyaystvennykh Meropriyatiy,* Leningrad, (1976).

Донченко, Р. В., Маркова, О. Л., Охинченко, А. И., Попов, И. В., Скальский, Б. Г. и Сорочан, О. Г. Исследования возможных последствий территориального перераспределения водных ресурсов, *Водные ресурсы,* № 1, (1977).

Donchenko, R. V., Markova, O. L., Okhinchenko, A. I., Popov, I. V., Skalskiy, B. G. and Sorochan, O. G. Studies of large-scale areal water resources redistribution, *Vodn. Resur.* No. 1, (1977).

Дроздов, О. А., Сорочан, О. Г., Маркова, О. Л., Охинченко, А. И., Бурова, Л. П. и Воскресенский, А. И. Возможные изменения элементов гидрологичесного цикла в северной части бассейна Оби под влиянием переброски стока, Сб. *Оценка возможных изменений режима низовьев и устьев рек Арктической зоны Западной Сибири под влиянием водохозяйственных мероприятий,* Ленинград, (1976).

Drozdov, O. A., Sorochan, O. G., Markova, O. L., Okhinchenko, A. I., Burova, L. P. and Voskresenskiy, A. I. Possible changes in the elements of the hydrological cycle in the northern part of the Ob Basin under the influence of streamflow transfer, *Otsenka Vozmozhnykh Izmeneniy Rezhima Nizovyev i Ustyev rek Arkticheskoy Zony Zapadnoy Sibiri pod Vliyaniyem Vodokhozyaystvennykh Meropriyatiy,* Leningrad, (1976).

Захаров, В. П. и Калачев, Н. С. Проблема территориального перераспрелеления водных ресурсов Срединного региона СССР, *Комплексное использование водных ресурсов Сибири и Дальнего Востока и их охрана,* Омск, (1969).

Zakharov, V. P. and Kalchev, N. S. The problem of water resources areal redistribution in the USSR Central Region, *Kompleksnoye Ispolzovaniye Vodnykh Resursov Sibiri i Dalnego Vostoka i ikh Okhrana,* Omsk, (1969).

Иванов, В. В. и Налимов, Ю. В. Состояние и перспективы изучения устьевых областей рек Западной Сибири для оценки возможных изменений их режима под влиянием водохозяйственных мероприятйи, *Труды Западно-Сибирского регионального научно-исследовательского гидрометеорологического института,* вып. 26, (1976).

Ivanov, V. V. and Nalimov, Yu. V. The present state and prospects of investigating the mouth areas of the West Siberian rivers to determine possible changes in their regimen under the influence of water management activities, *Tr. Zapadno-Sib. Reg. Naucho-Issled. Gidrometeorol. Inst.,* issue 26, (1976).

Иванов, М. С., Свашенко, Л. С. и Стольберг, Ф. В. Защита от истощения и регулирование качества поверхностных и подземных вод при осуществлении крупных межбассейновых перебросок, Сб. *Проблемы охраны и использования вод,* вып. 2, (1973).

Ivanov, M. S., Svashenko, L. S. and Stolberg, F. V. Quality control of surface and underground waters on transfer of large amounts of flow between basins, *Probl. Okhr. Ispolzovaniya Vod,* issue 2, (1973).

Иванов, М. С. и Стольберг, Ф. В. Вопросы охраны вод при межбассейновых перебросках стока, *Труды Всесоюзного совещания по охране вод,* Харьков (1973).

Ivanov, M. S. and Stolberg, F. V. The problems of water protection in case of inter-basin streamflow transfer, *Tr. Vses. Soveshch. Okhr. Vod,* Kharkov (1973).

Исмайылов, Г. Х. и Голубаш, Т. Ю. Учет межзональной и внутризональной асинхронности стока бассейнов рек Европейской территории Союза в связи с намечаемыми перебросками стока северных рек, Сб. *Комплексное использование водных ресурсов,* Минск (1975).

Ismaylov, G. Kh. and Golubash, T. Yu. Consideration of the inter-zonal and inside-zonal asynchronism in the flow from the river basins of the USSR European area in connection with the planned diversion of the flow of the northern rivers, *Kompleksnoye Ispolzovaniye Vodnykh Resursov,* Minsk, (1975).

Калинии, Г. П., Быков, В. Д., Евсеева, Л. С. и Попов, Л. И. Водные ресурсы СССР и их перераспределение для нужд народного хозяйства, Сб. *Взаимодействие поверхностного и подземного стока,* вып. 2, МГУ, Москва, (1974).

Kalinin, G. P., Bykov, V. D., Evseeva, L. S. and Popov, L. I. Water resources of the USSR and their reallocation for the needs of national economy, *Vzaimodeistvie Poverhnostnogo i Podzemnogo Stoka,* issue 2, Moscow State University, Moskva, (1974).

Карасев, И. Ф. *Русловые процессы при переброске стока,* Ленинград, (1970).

Karasev, I. F. *River Channel Processes on Streamflow Transfer,* Leningrad, (1970).

Карасев, И. Ф. Особенности русловых процессов при переброске стока, *Труды IУ Всесоюзного Гидрологического Съезда,* т. **10,** Ленинград (1976).

Karasev, I. F. The characteristics of river-bed processes in case of streamflow transfer, *Tr. Vses. Gidrol. S"êzda,* **10,** Leningrad (1976).

Комлев, А. М. и Вострякова, Н. В. Некоторые аспекты гидрологических проблем в связи с переброской части стока сибирских рек на юг, Сб. *Влияние перераспределения смока рек бассейна Оби на природу тайги Срединного региона,* Иркутск, (1975).

Komlev, A. M. and Vostryakova, N. V. Some aspects of the hydrological problems in connection with the diversion of part of the flow of the Siberian rivers southward, *Vliyaniye Pereraspredeleniya Stoke rek Basseyna Obi na Prirodu Taygi Sredinnogo Regiona,* Irkutsk, (1975).

Кузьмин, И. А. и Викулова, Л. И. Проблемы русловых процессов при переброске стока, *Водные ресурсы,* № 3, (1974).

Kuzmin, I. A. and Vikulova, L. I. The river-bed process problems in transferring river flow, *Vodn. Resur.* No. 3, (1974).

Львович, М. И. Географические аспекты территориального перераспределения водных ресурсов, *Изв. АН СССР, серия геграф.,* **№ 2,** (1977).

Lvovich, M. I. Geographical aspects of the large-scale reallocation of water resources, *Izv. AN SSSR, Ser. Geogr.,* No. 2, (1977).

Малик, Л. К. Об изменении некоторых элементов режима рек Нижней Оби при регулировании стока (в сявзи с проблемой перераспределения водных ресурсов в Срединном регионе), *Вестник МГУ, серия У – География,* № 1, (1975).

Malik, L. K. Variation of some elements of the regimen of the rivers of the Lower Ob area on streamflow regulation in connection with the water redistribution problem in the Central Region, *Vest. mosk. gos. Univ., ser. V-Geogr.,* 1, (1975).

Малик, Л. К. Гидрологические проблемы таежных районов Западной Сибири в связи с перераспределением водных ресурсов в Срединном регионе, Сб. *Влияние перераспределения стока рек бассейна Оби на природу тайги Срединного региона,* Иркутск, (1975).

Malik, L. K. Hydrological problems of the West Siberia taiga regions in connection with watei resources redistribution in the USSR central region, *Vliyaniye Pereraspredeleniya Stoka rek Basseyna Obi na Prirodu Taygi Sredinnogo Regiona,* Irkutsk, (1975).

Малик, Л. К. *Разработка модели оптимального режима рек переувлажненных зон Западной Сибири в связи с проблемами мелиорации территории и межзонального перераспределения водных ресурсов,* XXIII Международный Географический конгресс, Симпозиум "Преобразование водного баланса под влиянием хозяйственной деятельности", Ленинград, (1976).

Malik, L. K. *The development of an optimum regimen model for the rivers of the west Siberia's over-humid zones in connection with the soil improvement and inter-zonal water resources redistribution problems,* XXIII Mezhdunarodnyi Geograficheskiy Kongress, Simpozium, Preobrazovaniye Vodnogo Balansa pod Vliyaniem Khozyaystvennoy Deyatelnosti, Leningrad, (1976).

Малик, Л. К. Гидрологический аспект проблемы преобразования природы Западно-Сибирской равнины, *Водные ресурсы,* № 3, (1976).

Malik, L. K. The hydrological aspect of the environment remaking problem for the west Siberian plain, *Vodn. Resur.* No. 3, (1976).

Малик, Л. К. Изменение режима подпоров на притоках низовьев Оби при регулировании стока, Сб. *Оценка возможных изменений режима низовьев и устьев рек Арктической зоны Западной Сибири под влиянием водохозяйственных мероприятий,* Ленинград, (1976).

Malik, L. K. Change in backwater regimen on the Lower Ob Tributaries under runoff regulation conditions, *Otsenka Vozmozhnykh Izmeneniy Rezhima Nizovyev i Ustyev rek Arkticheskoy Zony Zapadnoy Sibiri pod Vliyaniem Vodokhozyaystvennykh Meropriyatiy,* Leningrand, (1976).

Маркова, О. Л., Охинченко, А. И. и Сорочан, О. Г. *Изменение водного баланса влагооборота и гидрологичесного режима рек под влиянием межзонального перераспределения водных ресурсов,* XXIII Международный Географический конгресс, Симпозиум "Преобразование водного баланса под влиянием хозяйственной деятельности", Ленинград, (1976).

Markova, O. L., Okhinchenko, A. I. and Sorochan, O. G. *Variation of the water balance of the water cycle and the hydrological conditions of rivers under the influence of inter-zonal water resources redistribution,* XXIII Mezhdunarodnyi Geograficheskii Kongress, Simpozium "Preobrazovaniye Vodnogo Balansa pod Vliyaniyem Khozyaystvennoy Deyatelnosti", Leningrad, (1976).

Марти, Ю. Ю. и Раткович, Д. Я. Водохозяйственные проблемы Азовского и Каспийского морей, *Водные ресурсы,* № 3, (1976).

Marti, Yu. Yu. and Ratkovich, D. Ja. Water management problems of the Sea of Azov and the Caspian Sea, *Vodn. Resur.* No. 3, (1976).

Мезенцев, В. С. О вероятных направлениях и размерах влияния перебросок части стока сибирских рек на гидролого-климатические условия лесной зоны Западной Сибири, Сб. *Природные условия Западной Сибири и переброска стока рек в Среднюю Азию,* Новосибирск, (1975).

Mezentsev, V. S. On possible impacts due to partial transfer to the flow of the Siberian rivers on hydrological and climatic conditions of the forest zone of West Siberia, *Prirodnye Usloviya Zapadnoy Sibiri i Perebroska Stoka rek v Srednyuyu Aziyu,* Novosibirsk, (1975).

Мезенцев, В. С. Вероятное влияние переброски части стока Оби и Иртыша на увлажненность лесной зоны, *Труды Омского сельскохозяйственного института,* **151,** (1976).

Mesentsev, V. S. The most likely effect of the partial transfer of the Ob and the Irtysh runoff on the precipitation–evaporation difference in the forest zone, *Tr. omsk. Sel'-khoz. Inst.,* **151,** (1976).

Мезенцев, В. С. О вероятных размерах изменения воднобалансовых элементов в зонах изъятия и распределения стока сибирских рек, *Труды Омского сельскохозяйственног института,* **151,** (1976).

Mezentsev, V. S. The most likely dimensions of changes in water balance components in the flow withdrawal and distribution zones of the Siberian rivers, *Tr. omsk. Sel'-khoz. Inst.* **151,** (1976).

Мезенцев, В. С. *О вероятных изменениях водного баланса Срединного региона СССР под влиянием межзонального перераспределения стока сибирских рек с целью гидротехнических мелиораций,* XXIII Мехдународный Географический конгресс, Симпозиум "Преобразование водного баланса под влиянием хозяйственной деятельности", Ленинград, (1976).

Mezentsev, V. S. *Possible changes in the water balance of the USSR central region under the influence of the inter-zonal transfer of the flow of the Siberian rivers for reclamation,* XXIII Mezhdunarodnyi Geograficheskiy Kongress, Simpozium, Preobrazovaniye Vodnogo Balansa pod Vliyaniyem Khozyaystvennoy Deyatelnosti, Leningrad, (1976).

Мезенцев, В. С. Вероятные изменения водного баланса под влиянием межзонального перераспределения стока сибирских рек, *Водные ресурсы,* № 1, (1977).

Mezentsev, V. S. Likely changes in the water balance under the influence of the inter-zonal redistribution of the Siberian river flow, *Vodn. Resur.,* No. 1, (1977).

Михайлов, В. Н., Рогов, М. М., Ган, Г. Н., Макарова, Т. А. и Полонский, В. Ф. Изменения гидролого–морфологических процессов в устьях рек под влиянием водохозяйственных мероприятий и нескоторые методы их расчета, Сб. *Оценка возможных изменений режима низовьев и устьев рек Арктической зоны Западной Сибири под влиянием водохозяйственных мероприятий,* Ленинград, (1976).

Mikhaylov, V. N., Rogov, M. M., Gan, G. N., Makarova, T. A. and Polonskiy, V. F. Variations of hydrological and morphological processes in the river mouth areas due to water utilization activities and some methods of estimating such variations, *Otsenka Vozmozhnykh Izmeneniy Rezhima Nizovyev i Ustyev rek Arkticheskoy Zony Zapadnoy Sibiri pod Vliyaniyem Vodokhozyaystvennykh Meropriyatiy,* Leningrad, (1976).

Николаев, И. И. Происходящне и ожидаемые изменения в экологии озёр Северо—Запада СССР, *Водные ресурсы,* № 6, (1977).

Nikolaev, I. I. Occurring and expected variations in the lake ecology of the USSR north west, *Vodn. Resur.* No. 6, (1977).

Новиков, С. М. Возможные изменения воднотеплового режима болот Северной части Западной Сибири под влиянием хозяйственной деятельности, Сб. *Оценка возможных изменений режима низовьев и устьев рек Арктической зоны Западной Сибири под влиянием водохозяйственных мероприятий,* Ленинград, (1976).

Novikov, S. M. Possible changes in the water and heat conditions of the marshes of the northern part of West Siberia under the influence of economic activity, *Otsenka Vozmozhnykh Izmeneniy Rezhimov Nizovyev i Ustyev rek Arkticheskoy Zony Zapadnoy Sibiri pod Vliyaniem Vodokhozyaystvennykh Meropriyatiy,* Leningrad, (1976).

Скорняков, В. А. Корроляционная связь колебаний годового стока рек Срединного региона, *Водные ресурсы,* № 5, (1975).

Skornyakov, V. A. Correlative relationship between annual variations in the river flow of the USSR central region, *Vodn. Resur.,* No. 5, (1975).

Скорняков, В. А. О возможном изменении высоты половодья и условий затопления поймы в нижних бьефах гидроузлов на Средней Оби и Нижнем Иртыше, *Водные ресурсы,* № 3, (1976).

Skornyakov, V. A. On possible changes in the flood level downstream hydraulic complexes at the Middle Ob and the Lower Irtysh, *Vodn. Resur.,* No. 3, (1976).

Чембарисов, Э. И. К решению гидрохимических задач в связи с переброской части стока Сибирских рек на юг страны, Сб. *Природно-мелиоративная характеристика Средней Азии и Казахстана,* Пущино, (1976).

Chembarisov, E. I. The solution of hydrochemical problems in connection with the diversion of part of the flow of the Siberian Rivers to the South of the USSR, *Prirodno-Meliorativnaya Kharakteristika Sredney Azii i Kazakhstana,* Pushchino, (1976).

Черненко, И. М. Переброска части стока сибирских рек на юг и оптимальные условия сохранения Арала, *Проблемы освоения пустынь,* № 6, (1972).

Chernenko, I. M. Partial diversion of the flow of the Siberian Rivers to the South and the optimum conditions to protect the Aral Sea, *Probl. Osvoeniya Pustyn,* No. 6, (1972).

Широков, В. М. Перераспределение твердого стока Оби после переброски части её вод в Среднюю Азию, Сб. *Оценка возможных изменений режима низовьев и устьев рек Арктической зоны Западной Сибири под влиянием водохозяйственных мероприятий,* Ленинград, (1976).

Shirokov, V. M. Change in sediment transport in Ob River due to partial transfer of its runoff to Central Asia, *Otsenka Vozmozhnykh Izmeneniy Rezhima Nizovyev i Ustyev rek Arkticheskoy Zony Zapadnoy Sibiri pod Vliyaniem Vodokhozyaystvennykh Meropriyatiy,* Leningrad, (1976).

Шпайхер, А. О., Морецкий, В. Н. и Никифоров, Е. Г. Океанографический режим Северного Ледовитого океана и проблема перераспределения стока рек его бассейна, Сб. *Оценка возможных изменений режима низовьев и устьев рек Арктической зоны Западной Сибири под влиянием Водохозяйственных мероприятий,* Ленинград, (1976).

Shpaykher, A. O., Moretskiy, V. N. and Nikiforov, E. G. The oceanographical regimen of the Arctic Ocean and the problem of river flow redistribution in its basin, *Otsenka Vozmozhnykh Izmeneniy Rezhima Nizovyev i Ustyev rek Arkticheskoy Zony Zapadnoy Sibiri pod Vliyaniem Vodokhozyaystvennykh Meropriyatiy,* Leninegard, (1976).

TECHNOLOGY

Anon., *Operating principles: Frying pan-Arkensas project,* 87th Congr., 1st Session, House Doc. 130, 7 pp. (1961).

Burkholder, J. L. Colorado River aqueduct tunnels, *Civ. Engng,* **5,** 524–527, Disc: Agnew, J. Carson, **5,** 643 (1935).

Christensen, N. A. *et al., Economic analysis of large water systems,* Cornell Univ. Water Resources Publn, No. 22, 98 pp. (1969).

Christensen, B. A. and Bush, Peter W. Optimum design of inter-basin tunnel systems, *Water Resour. Bull.,* **7,** 273–284 (1971).

Dann, H. E., Hartwig, W. P. and Hunter, J. R. Unlined tunnels of the Snowy Mountains Hydro-Electric Authority, Australia, *Proc. Am. Soc. civ. Engrs,* **90** (PO3), 47–49, Disc: **91** (PO1), 118–124; **92** (PO1), 125–129 (1964).

Delord, Pierre, Quelques ouvrages originaux du canal principal d'irrigation du Bas Rhone– Languedoc (Some Innovations of the Main Irrigation Canal of the Bas Rhone-Languedoc Project), *Travaux,* **49,** 87–96 (1965).

Golze, Alfred R. Use of pipelines in the California Water Project, *Proc. Am. Soc. civ. Engrs,* **93** (PL-3), 25–32, Disc: **94** (PL-1), 198–199 (1967).

Gumensky, D. B. The Colorado River Aqueduct: Governing factors in aqueduct design, *Engng News Rec.* **121,** 653–658 (1938).

Halwani, Salah, Janod, Albert, and Lajeat, J. P. Le percement du tunnel d'Avali de l'amenagement du Litani au Liban (The driving of the Awali Tunnel of the development of the Litani in Lebanon), *Travaux,* **49,** 247–260 (1965).

Hinds, Julian, Large delivery lines for the Colorado River Aqueduct System, *Civ. Engng,* **8,** 469–472 (1938).

Hinds, Julian, The Colorado River Aqueduct, major problems of aqueduct location, *Engng News Rec.,* **121,** 646–652 (1838).

Howson, Louis R. Factors affecting long-distance transmission of water, *J. Am. Wat. Wks Ass.*, **49,** 1359–1368 (1957).

Hunter, J. R. and Hartwig, W. P. The design and construction of the Tooma–Tumut project of the Snowy Mountains Scheme, *J. Instn Engrs Aust.*, **34,** 163–185 (1962).

Klingeman, Peter C. *Engineering aspects of diversion proposals,* Symposium on Northwest–Southwest Water Diversion Issues, Oregon State Univ., pp. 11–30 (1967).

Linaweaver, F. P. Jr and Clark, C. Scott, Costs of water transmission, *J. Am. Wat. Wks Ass.*, **56,** 1549–1560 (1964).

Miller, David R. and Miller, James P. Optimizing conveyance systems for water transmission, (1) Am. Soc. Civ. Engrs National Convention, Meeting reprint 603, San Diego, CA, 28 pp; (2) *Civ. Engng,* **38,** 38–41 (July 1968).

Resnick, Sol, Large-scale transfers of water and alternatives, in McGinnes and Goldman, *Arid Lands in Perspective,* U. of Arizona Press, (1968) pp. 418–421.

Reynolds, Robin R. and Madsen, Welby R. Automation in California's State Water Project, *Proc. Am. Soc. civ. Engrs,* **93** (PL-3), 15–23, Disc: **94** (PL-1), 196–198 (1967).

Socha, Max K. Construction of the second Los Angeles aqueduct, *J. Am. Wat. Wks Ass.*, **57,** 699–706 1965).

Tinney, E. R. Engineering aspects (of NAWAPA), *Bull. atom. Scient.*, **23,** 21–25 (September 1967).

Weymouth, F. E. Tunnel construction—Colorado River Aqueduct, *Civ. Engng,* **4,** 240–244 (1934).

Wineland, J. A. and Lucas, C. V. *Economic comparison of canal linings for the California Aqueduct,* Proc. Seepage Symposium, Phoenix, Arizona, USDA-ARS, pp. 41–90, (19–21 February 1963).

Young, Gale, *Import alternatives, reuse, desalination and rain-making are discussed,* AAAS Symposium on Arid Lands, Dallas, pp. 382–397 (31 December 1968).

Альхименко, А. И., Ермаков, В. С. и Симаков, Г. В. О воздействии ветровых волн на откосы главного канала переброски, *Труды Ленинградского Политехнического института,* № **351,** (1976).

Alkhimenko, A. I., Ermakov, V. S. and Simakov, G. V. The effects of wind waves on the slopes of the main diversion canal, *Tr. Leningr. Politekh. Inst.*, No. **351,** (1976).

Бусалаев, И. В., Павленко, В. П. и Сергеева, Л. В. Экономико–математическая модель оптимизации схемы преодоления водораздела при межбассейновых перебросках стока, Сб. *Проблемы гидроэнергетики и водного хозяйства,* вып. 10, (1973).

Busalaev, I. V., Pavlenko, V. P. and Sergeeva, L. V. An economic and mathematical model of optimization for watershed divide passing schemes in case of inter-basin flow transfers, *Probl. Gidroenerg. Vodn. Khoz.*, Issue 10, (1973).

Васильев, Ю. С., Лойцянская, И. Л. и Шарыгин, В. С. Исследования эффективности насосных станций с обратимыми агрегатами при региональной переброске стока, *Труды Ленинградского Политехнического института,* № **351,** (1976).

Vasilyev, Yu. S., Loytsyanskaya, I. L. and Sharygin, V. S. Studies of the effectiveness of reversible-unit pumping stations during interregional flow transfer, *Tr. Leningr. Politekh. Inst.*, No. **351,** (1976).

Дмитриев, Г. С. К вопросу о переброске стока при гидроэнергетическом стороительстве, Сб. *Вопросы энергетики Кольского полуострова,* Апатиты, (1975).

Dmitriev, G. S. On the problem of streamflow transfer in connection with hydropower construction, *Voprosy Energetiki Kolskogo Poluostrova,* Apatity, (1975).

Замахаев, В. С., Мизинов, П. И. и Шкундин, Б. М. Пути технического прогресса в крупном гидротехническом строительстве при межзональном перераспределении стока, *Труды Гидропроекта,* сб. 42, (1974).

Zamakhaev, V. S., Mizinov, P. I. and Shkundin, B. M. Methods for engineering improvement of large-scale hydraulic construction in inter-zonal flow redistribution, *Tr. Gidroproekta,* issue 42, (1974).

Смолко, С. Я. и Ялыгин, В. М. Об исследовании высокоскоростных судов на трассе главного канала переброски, *Труды Ленинградского Политехнического института,* № **351,** (1976).

Smolko, S. Ya. and Yalygin, V. M. Studies of high-speed vessels on the route of the main diversion canal, *Tr. Leningr. Politekh. Inst.*, No. **351,** (1976).

Стольберг, Ф. В. Пути снижения стоимости инженерных комплексов для межбассейновой переброски стока, Сб. *Проблемы охраны и использования вод,* вып. 3, Харьков (1973).

Stolberg, F. V. Methods for reducing the costs of engineering complexes for inter-basin streamflow transfer, *Probl. Okhr. Ilspolzovaniya Vod,* issue 3, Kharkov (1973).

MODELLING AND WATER MANAGEMENT

Bayley, Fred H. III, *The role of the Mississippi River Commission in the Mississippi River Water Export Study,* Symposium on Aspects of the Diversion of Mississippi River Water to Texas and New Mexico, pp. 77–94 (1969).

Boswell, Howard B. *Appendix I: The Texas Water Plan,* Symposium on Aspects of the Diversion of Mississippi River Water to Texas and New Mexico, pp. 136–154 (1969).

Burleigh, Harry P. *The Bureau of Reclamation and the Mississippi River Export Plan,* Symposium on Aspects of the Diversion of Mississippi River Water to Texas and New Mexico, pp. 95–106 (1969).

Fisher, Anthony C. *Some theoretical and measurement issues in economic-assessment of inter-basin water transfers,* IIASA Task Force Meeting on Interregional Water Transfers, Laxenburg, (11–14 October 1977).

Harwuna, M. Okado N. *System analysis for the planning of facilities in inter-basin water distribution problems,* Hemairs Faculty of Engineering, Kyoto Univ., Japan, No. 3, (1974).

Leonard, Robert L. *Integrated management of ground and surface water in relation to water importation,* Giannini Foundation of Agricultural Economics Res. Rep. No. 279, Univ. of California, Berkeley, (1964).

Lovell, Troy Lynn, *A comparison of system designs and operations for the transmission of large volumes of water to West Texas,* Thesis for the M.S. in C.E. Degree, Texas Tech., Lubbock, 68 pp. (1969).

McDermott, Alex. D. *Western states water planning,* Symposium on Aspects of the Diversion of Mississippi River Water to Texas and New Mexico, pp. 59–76 (1969).

Meier, W. L., Weiss, A. O. and Puentes, C. D. *Optimum water management under the Texas Water Plan,* Proc. 14th Congr. Int. Ass. for Hydraulic Res., 5, pp. 351–354 (1971).

Miller, David R. and Miller, James P. Optimizing conveyance systems for water transmission, (1) Am. Soc. Civ. Engrs National Convention, Meeting Reprint 603, San Diego, CA, 28 pp; (2) *Civ. Engng,* **38,** 38–41 (July 1968).

Orlob, G. T. *Planning the Texas Water System,* Fifth Ann. Water Resources Res. Conf., Washington DC, 12 pp., Proc., (3–4 February 1970).

Reynolds, S. E., *Water development and planning for New Mexico,* Symposium on Aspects of the Diversion of Mississippi River Water to Texas and New Mexico, pp. 38–58 (1969).

van der Riet, P. *Simulation of inter-basin water transfer systems with special reference to the Tugela–Vaal Scheme,* Univ. of Witwatersrand Hydrological Research Unit Rep. 8/75, 100 pp. (1975).

Skugerboe, Gaylord V. and Huntzinger, Thomas L. Effects of inter-basin transfers upon water management alternatives in Central Utah, *Water Resour. Bull.,* **7,** 1027–1037 (1971).

Symposium on Aspects of the Diversion of Mississippi River Water to Texas and New Mexico; Panel Discussion, pp. 107–135 (1969).

Tinney, E. Roy, *Planning for massive water resource development,* Symposium on The Continental Use of Arctic-Flowing Rivers, Washington State Univ., pp. 165–176 (1968).

Whetstone, George A. Utilization and control of rivers in the Southwest, *Rivers* '76, Am. Soc. Civ. Engrs, pp. 1149–1158 (1976).

Абрамов, А. В., Озиранский, С. Л. и Раппопорт, В. В. Оптимизация комплексного использования водных ресурсов межбассейновых перебросок стока, Сб. *Комплексное использование водных ресурсов, Тезисы докладов Всесоюзного научно-технического совещания,* Минск, (1975).

Abramov, A. V., Oziranskiy, S. L. and Rappoport, V. V. The optimization of the integrated utilization of the water resources of the flow redistributed between basins, *Kompleksnoye Ispol zovaniye Vodnykh Resursov, Tezisy Dokladoy Vsesoyuznogo Nauchnotechnicheskogo Soveschaniya,* Minsk, (1975).

Агарков, С. Г., Дружинин, И. П., Коноваленко, З. П., Кузнецов, А. Н., Наумов, Т. С., Ушакова, В. В., Герарди, И. А., Дорфман, Н. Л., Коган, Р. Н., Кокорев, Е. П. и Рыскулов, Д. М. Оптимизация размещения водопотребляющих производств, (На примере сельского хозяйства страны), *Водные ресурсы,* № 3, (1977).

Agarkov, S. G., Druzhinin, I. P., Konovalenko, Z. P., Kuznetsov, A. N., Naumova, T. S., Ushakova, V. V., Gerardi, I. A., Dorfman, N. L., Kogan, R. N., Kokorev, E. P. and Ryskulov, D. M. The optimization of the distribution of water-consuming enterprises in the USSR agriculture, *Vodn. Resur.* No. 3, (1977).

Волков, И. А. Некоторые вопросы преобразования водного хозяйства Сибири и решениепроблемы переброски, Сб. *Природные условия Западной Сибири и переброска стока рек в Среднюю Азию,* Новосибирск, (1975).

Volkov, I. A. Some questions of the transformation of Siberia's water resources and the solution of the diversion problem, *Prirodnye Usloviya Zapadnoy Sibiri i Perebroska Stoka rek v Srednyuyu Aziyu,* Novosibirsk, (1975).

Воропаев, Г. В. Проблема перераспределения водных ресурсов страны и её моделирование, *Гидротехника и мелиорация,* № 9, (1976).

Voropaev, G. V. Reallocation of national water resources and its modelling, *Gidrotekh. Melior.* No. 9, (1976).

Воропаев, Г. В. Единая водохозяйственная система страны, *Водные ресурсы,* № 6, (1976).

Voropaev, G. V. Joint water management system of the country, *Vodn. Resur.* No. 6, (1976).

Воропаев, Г. В. и Киселёв, В. Г. К вопросу об оценке эффективности проектных схем перераспределения водных ресурсов, *Методы системного анализа в проблемах рационального использования водных ресурсов,* т. II, Москва (1976).

Voropaev, G. V. and Kiselev, V. G. Evaluating the effectiveness of draft schemes for water resources redistribution, *Metody Sistemnogo Analiza v Problemakh Ratsionalnogo Ispolzovaniya Vodnykh Resursov,* Vol. II, Moskva (1976).

Дунин-Барковский, Л. В. и Моисеев, Н. Н. Система моделей перераспределения речного стока СССР, *Водные ресурсы,* № 3, (1976).

Dunin-Barkovskiy, L. V. and Moiseev, N. N. A system of models for river run-off redistribution in the USSR, *Vodn. Resur.* No. 3, (1976).

Замахаев, В. С. О разработке научных основ к проектным проработкам по комплексному использованию и охране водных ресурсов СССР на период до 2000 года, *Труды Гидропроекта,* сб. 26, (1972).

Zamakhaev, V. S. The development of scientific principles for draft schemes of water resources integrated utilization and protection in the USSR for the period up to 2000, *Tr. Gidroproekta,* issue 26, (1972).

Маркиш, М. С. Об оптимальном управлении уровнем бессточного водоёма (на примере Каспийского моря), *Водные ресурсы,* № 5, (1977).

Markish, M. S. On the optimal control of the water level of a closed water body, exemplified by the Caspian Sea, *Vodn. Resur.,* No. 5, (1977).

Окороков, В. Р., Фёдоров, М. П., Шапот, Д. В. и Щавелёв, Д. С. Исследование эффективности регионального перераспределения речного стока на основе многокритериального анализа, *Труды Ленинградского Политехнического института,* № **351,** (1976).

Okorokov, V. R., Fedorov, M. P., Shapot, D. V. and Shchavelev, D. S. Studying the effectiveness of regional river flow redistribution on the basis of a multi-criterion analysis, *Tr. Leningr. Politekh. Inst.* No. **351,** (Leningrad), 1976.

Шаломаев, М. И. Проблемы водообеспечения республик Средней Азии и Казахстана, *Гидротехника и мелиорация,* № 9, (1975).

Shalomaev, M. I. The problems of water supply in central Asia and Kazakhstan, *Gidrotekh. Melior.,* No. 9, (1975).

JURIDICIAL AND POLITICAL ISSUES

Anon., *Operating principles: Fryingpan–Arkansas Project,* 87th Congr., 1st Session, House Doc. 130, 7 pp. (1961).

Bessey, Roy F. *Political implications,* Symposium on Northwest–Southwest Water Diversion Issues, Oregon State Univ., pp. 59–77 (1967).

Beveridge, James A. *The export of surplus water from Canada to the United States,* Cornell Univ., Water Resources Center, Pbln No. 11, pp. 16–29 (1965).

Bourne, Charles B. *International law and the diversion of the Columbia River in Canada,* Univ. of British Columbia Lecture Series No. 27, pp. 17–25 (1956).

Clark, Chapin D. *Issues and legal aspects of diversion,* Symposium on Northwest–Southwest Water Diversion Issues, Oregon State Univ., pp. 31–46 (1967).

Dersch, Eckhart, *Water resources development—the river basin and the multi-basin region,* Cornell Univ., Water Resources Center, Pbln No. 11, pp. 1–15 (1965).

Ellis, Willis H. Water transfer problems: Law, in Kneese, Allen V. and Smith, Stephen C. (eds) *Water Research,* Johns Hopkins Press, Baltimore, for Resources for the Future Inc., (1966).

Engelbert, Ernest A. Policy issues of the Pacific Southwest water plan, *West. Resour. Pap.,* **6,** 157–204 (1964).

Fox, Irving K. *Some political aspects of large-scale inter-basin water transfers (summary),* AAAS Symposium on Arid Lands, Dallas, Texas, 6 pp. processed, (30 December 1968).

Frederiksen, Harald D. Design of California Aqueduct, *Proc. Am. Soc. civ. Engrs,* **95** (IR-2), 307–328 (1969).

Hartman, L. S. and Seastone, Don, *Water Transfers: Economic Efficiency and Alternative Institutions,* Resources for the Future, Baltimore, (1970) 127 pp.

Johnson, Corwin W. and Knippa, Larry D. Transbasin diversion of water, *Texas Law Rev.,* **43,** 1035–1061 (1965).

Johnson, Ralph W. *Major inter-basin transfers—legal aspects,* Legal Study No. 7, National Water Commission, 206 pp. (1971).

Krieger, James H. *Importance of recognition of paramount rights in areas of origin,* Proc. Western Interstate Water Conf., pp. 110–116 (1964).

Ladner, Leon J. *Diversion of Columbia River Waters,* Univ. of British Columbia Lecture Series No. 27, pp. 1–16 (1956).

Laycock, Arleigh H. Canadian water for Texas? *Water Resour. Bull.,* **6,** 542–549 (1969).

Laycock, Arleigh H. Inter-basin transfer—the international dimension, *Water Resour. Bull.,* **7,** 1017–1026 (1971).

Mann, Dean E. *Inter-basin water transfers: A political and institutional analysis,* Natl. Water Comm. Publ. NWC-SBS-72-037, 157 pp. (1972).

O'Leary, Dillon, Canada guards its water against our big thirst, *Reporter,* **34,** 36–38 (27 January 1966), Disc: **34,** p. 13 (24 February 1966).

Piper, Don C. International law as environment for municipal litigation: The Chicago diversion cases, *Am. J. Int. Law,* **62,** 451–454 (1968).

Quinn, Frank J. *Area-of-origin protectionism in western waters,* Doctoral Dissertation, Univ. of Washington, 278 pp. +plates, (1970).

Quinn, Frank J. *Area-of-origin protectionism in western waters,* Social Science Series No. 6, Inland Waters Directorate, Water Planning and Management Branch, Ottawa, Canada, (1973).

Seastone, D. A. and Hartman, L. M. Alternative institutions for water transfers: The experience in Colorado and New Mexico, *Land Econ.,* **39,** 31–42 (1963).

Simsarian, James The diversion of waters affecting the United States and Canada, *Am. J. Int. Law.,* **32,** 488–518 (1938).

Simsarian, James, The diversion of interstate waters in the United States, *Am. polit. Sci. Rev.,* **32,** 907–921 (1938).

Simsarian, James, The diversion of waters affecting the United States and Mexico, *Texas Law Rev.,* **17,** 27–61 (1938).

Snel, J. Problemes d'aujourd'hui et de demain: la gestion de l'eau dans le domaine international (Problems of today and tomorrow: Water management in the international domain), *Techq. Eau Assain.,* **23,** 15–24 (Mars 1969).

Taylor, E. F. Arizona v. California—Landmark on the Colorado River, *Groundwater,* **5** (2), 5–12 (1967).

Towner, P. A. Protection for areas of origin and areas of export in water projects: The California Experience, *Water for Peace,* **5,** 637–648 (1967).

Tyler, Roger, *Water law in terms of planning a future for West Texas,* Proc. 6th West Texas Conf., pp. 89–122 (1968).

Weinberg, Edward, *Intrastate, interstate and international legal and administrative problems of large-scale water transfer,* AAAS Symposium on Arid Lands, Dallas, Texas, processed pp. 352–358 (31 December 1968).

Whetstone, George A. *The changing attitude in inter-basin transfer of water,* Proc. 7th West Texas Conf,. pp. 40–45 (1969).

Whetstone, George A. *A Critique of Ralph W. Johnson's "Major Inter-basin Transfers—Legal Aspects",* Water Res. Center, Texas Tech., Pbln. WRC 72-1, 25 pp. (1972).

Index